Operational Aerospace Medicine: A Flight Surgeon's Comprehensive Guide

by

Brian N. Karlsson

Published by Oryson Press
6605 Longshore Street
Dublin, OH 43017
E: info@orysonpress.com
W: www.orysonpress.com

First Edition 2025

Operational Aerospace Medicine: A Flight Surgeon's Comprehensive Guide
Brian N. Karlsson

Library of Congress Cataloging-in-Publication Data Available
British Library Cataloguing in Publication Information Available

ISBN 979-8-88715-471-8 (Hardback)

Published in the United States of America

TABLE OF CONTENTS

Chapter 7 Emergency Egress from Aircraft: Procedures and Equipment

Chapter 8 Pharmacological Considerations for Aircrew Members

Chapter 9 Fatigue Management in Aerospace Operations

Chapter 10 Psychiatric Health of Flying Personnel

Chapter 11 Aircraft Toxicology: Managing Hazardous Gases and Vapors

Chapter 12 Nutritional Guidelines for Aircrew and Passengers

Chapter 13 Aeromedical Support for Rescue and Survival Operations

Chapter 1: Optimizing Aircrew Performance and Well-being

The pursuit of air superiority and the advancement of aeronautical technology have consistently presented formidable human challenges. Aviation medicine, as a specialized scientific discipline, has fundamentally recognized the nearly insurmountable problems imposed upon the human physiological and psychological systems by modern military aircraft. A dedicated cadre of research scientists, representing diverse disciplines within aviation medicine, is perpetually engaged in endeavors aimed at enabling human beings to effectively adapt to the extreme conditions inherent in advanced aviation. These conditions include the escalating speeds, increasingly higher altitudes, extended operational ranges, and the heightened operational complexity that characterize both current and future aircraft designs.

Throughout its evolution, aeromedical research has achieved a substantial degree of success in the discovery and development of specialized equipment and intricate procedures. These innovations are crucial for empowering military aviators to maintain pace with the rapid advancements in aeronautical science and operational doctrine. However, the ultimate value and impact of these extensive aeromedical research efforts are intrinsically linked to the successful application of their findings by the individual Flight Surgeon. Therefore, the art and practice of aviation medicine must be vigorously pursued by every Flight Surgeon, demanding full utilization of all available knowledge and resources. The principal objective driving this comprehensive activity is the continuous maintenance of the flyer in the highest possible state of effectiveness and readiness under all conceivable circumstances. This paramount goal is best achieved through the meticulous planning and execution of a comprehensive aircrew effectiveness program, as outlined in official directives such as AFR 160-69. The successful fulfillment of such a program necessitates the diligent accumulation of both general and specific knowledge pertaining to the multifaceted problems confronting flyers, coupled with the strategic and full utilization of all available resources in the effective solution of these problems.

The scope of the aircrew effectiveness program is remarkably broad, encompassing the complete application of the full spectrum of aviation medicine's resources to meet the operational requirements of the flyer. The

successful prosecution of this program typically resolves into a systematic process: first, the accurate recognition of the problems faced by the flyer, and second, the proper and timely deployment of available means and knowledge for their resolution. To underscore the sheer multiplicity and intricate complexity of the challenges that arise, it is pertinent to briefly enumerate and describe a few broad categories of aircrew effectiveness problems, and concurrently indicate the established means available for their amelioration and solution.

Medical Selection and Examination Procedures

The initial and enduring mission of aviation medicine remains the rigorous selection of flying personnel. This activity has become progressively more critical with the advent and proliferation of supersonic jet flight, which places unprecedented physiological and psychological demands on aircrew. The Flight Surgeon's contribution to this selection process is therefore crucial, resting upon the accurate and expert accomplishment of prescribed examining procedures.

These procedures are meticulously applied in accordance with established medical standards for flying duties. It is important to note that the detailed procedures for medical examination are no longer contained within this specific edition of the Flight Surgeon's Manual. Instead, they have been fully integrated into the comprehensive Manual of Medical Examination (AFM 160-1). This dedicated manual encompasses all current medical standards, delineates the physical profile serial system, and provides detailed examining techniques, thereby ensuring convenience for frequent correlative reference and consistent application across all evaluations.

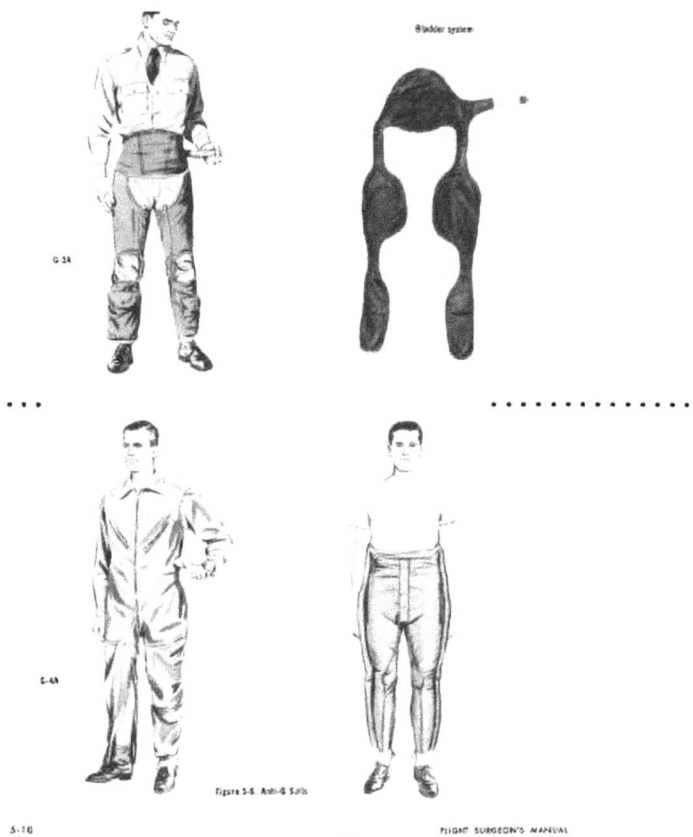

Managing Flight-Induced Abnormalities

Beyond initial selection, the Flight Surgeon is tasked with the ongoing management of medical conditions that may either be induced by the unique stresses of flight or may otherwise affect an individual's ability to fly safely and effectively. These challenges necessitate the application of the finest diagnostic abilities and the exercise of the best possible medical judgment. A medical condition may manifest as acutely or chronically induced by the inherent stresses of flying, or it may arise from other, unrelated causes. In any event, a precise diagnosis is imperative, followed by proper and timely treatment, and a thorough evaluation of its impact on the individual's flying status.

In managing such cases, the Flight Surgeon must maintain adequate administrative control over the flyer to prevent any untoward incidents that might occur if an unfit individual were to fly. This administrative oversight ensures that temporary or permanent conditions that could compromise flight safety are addressed appropriately. The foundation for successful management in this category lies in the provision of good medical care, close and continuous observation of aircrew health, and the proper execution of the periodic medical examination program. These combined efforts are designed to effectively address the myriad problems that can arise from flight-induced abnormalities and medical conditions.

For example, among the various flight-induced abnormalities, changes in barometric pressure can lead to distinct medical conditions. Aerotitis media, an acute or chronic traumatic inflammation of the middle ear, arises from pressure differences between the middle ear and the surrounding atmosphere. This condition, characterized by pain, deafness, tinnitus, and occasionally vertigo, necessitates prompt recognition and management. Similarly, aerosinusitis, an acute or chronic inflammation of the nasal accessory sinuses, results from similar pressure differentials, typically manifesting as severe pain in the affected region. The Flight Surgeon's expertise is crucial in distinguishing these from other conditions and initiating appropriate intervention, including techniques such as politzerization or, in refractory cases, catheterization, and even paracentesis if necessary, to restore physiological function. Delay in addressing such issues can not only cause significant discomfort and temporary incapacitation but also lead to chronic problems affecting an aircrew member's long-term flying capability.

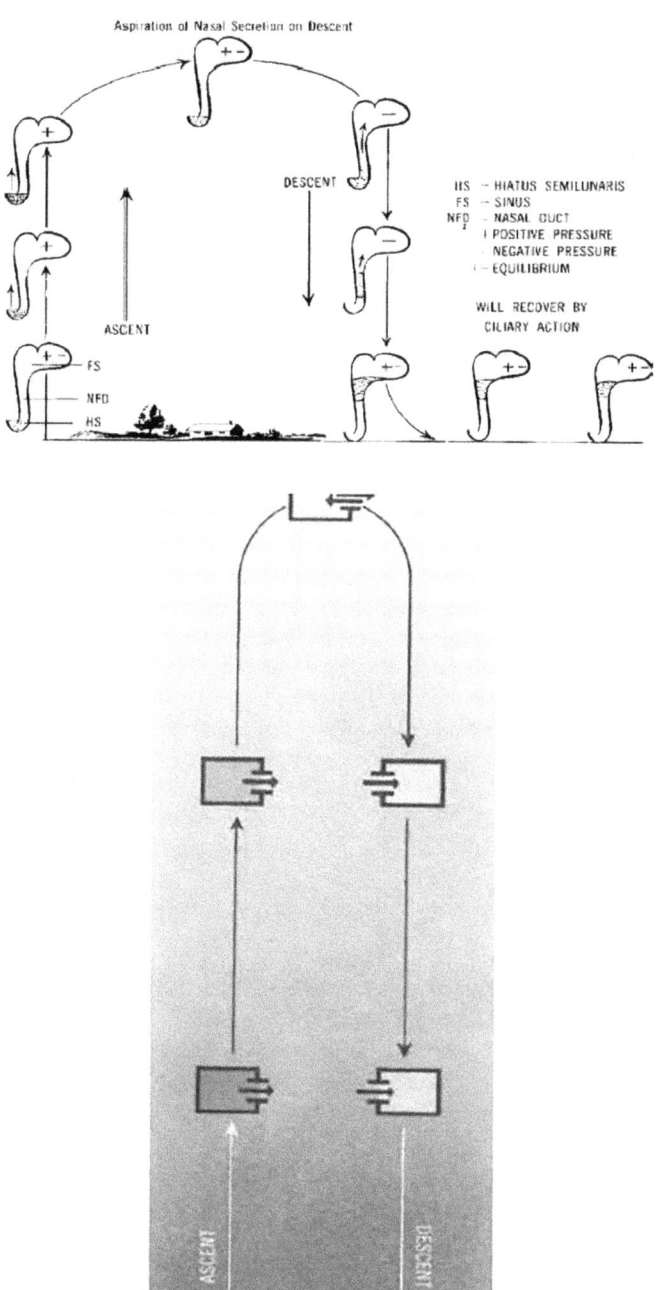

Furthermore, infections of the upper respiratory tract, such as acute otitis media, while not directly flight-induced, can significantly impact an aircrew member's ability to equalize pressure, predisposing them to aerotitis media during flight. Therefore, the Flight Surgeon's mandate includes vigilant screening and appropriate treatment of such infections, often involving antibiotics, to prevent flight-related complications. The principle of conservative management, followed by more intensive procedures when indicated, is consistently applied to ensure the flyer's rapid and safe return to duty. The ability to identify underlying issues, such as allergic rhinitis or anatomical obstructions, that might compromise Eustachian tube function is also paramount, as these predisposing factors must be addressed to prevent recurrent flight-induced barotrauma.

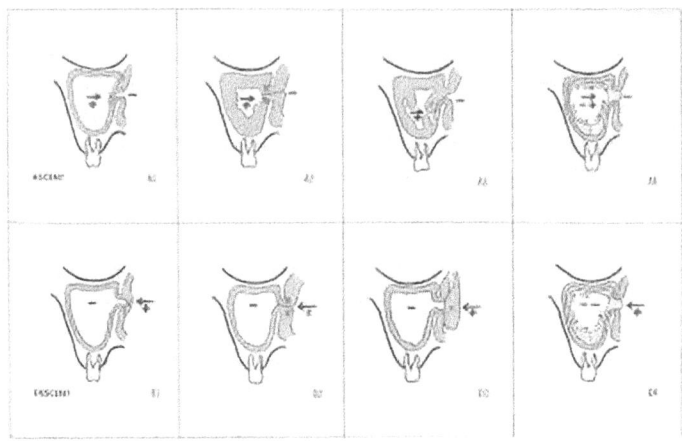

Protecting Against Flighlt Hazards (Hypoxia, Dysbarism, Accelerative Forces, Noise, Toxins, etc.)

One of the most critical problem areas encountered by the Flight Surgeon involves the protection of the flyer against the diverse hazards and inherent stresses of flight. The complexity of these problems is readily recognized when considering the multitude of environmental challenges faced by aircrew inmodern operational aircraft. These hazards stem from extreme physiological conditions and require sophisticated aeromedical interventions.

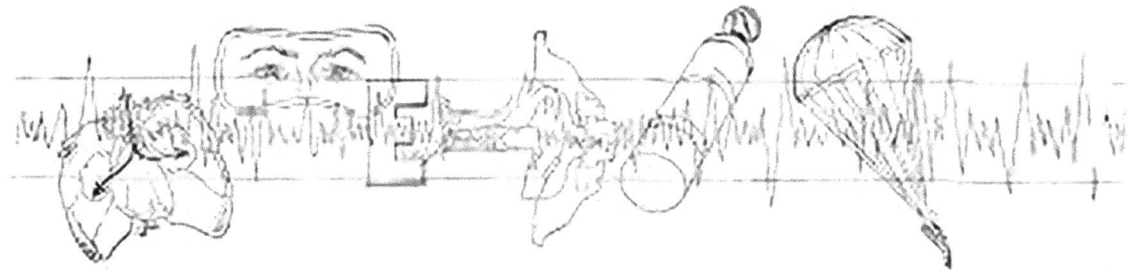

Hypoxia: High altitude flight introduces several significant hazards, with hypoxia being paramount. Hypoxia, or altitude sickness, is a syndrome typically acute, resulting from inadequate oxygenation of tissues due to a decreased partial pressure of oxygen in the inspired air. While sometimes referred to as anoxia, the term hypoxia (meaning deficiency rather than complete lack of oxygen) is more accurate, as tissues are rarely entirely devoid of oxygen even in severe cases. The symptomatology of acute hypoxic hypoxia depends on factors such as absolute altitude, rate of ascent, duration at altitude, ambient temperature, physical activity, and individual tolerance. Symptoms progress through indifferent, compensatory, disturbance, and critical stages, ranging from impaired dark adaptation and increased pulse rate in the indifferent stage, to fatigue, dizziness, and intellectual impairment in the disturbance stage, culminating in unconsciousness in the critical stage. The Flight Surgeon's role involves prophylactic measures such as 100% oxygen administration and training in oxygen equipment use. Oxygen systems, from continuous flow to demand-type and pressure-breathing systems, are designed to combat hypoxia, necessitating constant development and proper application. Liquid oxygen (LOX) systems represent an advanced solution for extended flight durations by providing a compact, lightweight source of breathing oxygen.

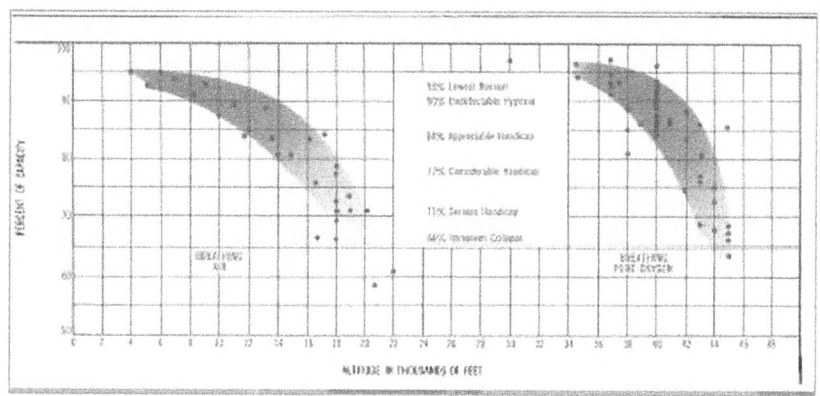

Dysbarism: Another crucial high-altitude hazard is dysbarism, a general term encompassing physiological effects from reduced barometric pressure, independent of hypoxia. This condition, often termed decompression sickness, manifests through various symptoms, broadly classified by their etiology: those from evolved gases (e.g., bends, chokes, neurological disorders) and those from the expansion of trapped gases (e.g., abdominal gas pain, aerodontalgia). Bends, characterized by deep, migratory pain in the extremities, and chokes, presenting with substernal burning and nonproductive cough, are common manifestations of evolved gas bubbles. Abdominal pain, resulting from expanding gases in the gastrointestinal tract, and aerodontalgia, pain in the jaw and teeth due to trapped gas, are other significant concerns. Factors influencing dysbarism include altitude, rate of ascent, duration of exposure, activity, cold, and individual susceptibility. The primary prophylactic measure against bends and chokes is denitrogenation, achieved by breathing reduced partial pressure nitrogen air or pure oxygen before ascent. Treatment primarily involves recompression to ground level, with severe cases requiring vigorous supportive measures for neurocirculatory collapse. The Flight Surgeon's vigilance is essential in preventing and managing these conditions, which can be exacerbated by rapid decompression events in pressurized aircraft.

Accelerative Forces: Modern military aircraft frequently subject aircrew to significant accelerative forces, broadly categorized as linear, angular, and radial. These forces arise from changes in velocity or direction and can have profound physiological effects. Linear acceleration occurs during changes in speed in a straight line (e.g., catapult takeoffs or crash landings), while radial acceleration results from changes in direction (e.g., tight turns or pulling out of a dive). Angular acceleration, a combination of both, is seen in maneuvers like spins. The effects of these forces depend on their intensity, duration, rate of application, the area and site of application on the body, and critically, their direction relative to the body's long axis.

- Positive G-forces (head-to-foot direction) cause blood displacement caudally, leading to a progressive sequence of visual dimming, blackout (loss of peripheral vision), and ultimately unconsciousness as cerebral blood flow is compromised. Pilots employ techniques like the M-1 maneuver (straining and muscle tensing) and wear anti-G suits to counteract these effects, increasing their G-tolerance.

- **Negative G-forces** (foot-to-head direction) result in increased arterial pressure in the head, potentially causing "red out" (vision obscured by engorged eyelid vessels) and discomfort, though severe intracranial hemorrhage is rare due to cerebrospinal fluid buffering.

- **Transverse G-forces** (acting perpendicular to the body's long axis, as in a prone or supine position) are the most tolerable, as they interfere minimally with blood flow. However, extreme values can still cause organ displacement, respiratory difficulties, and chest pain.

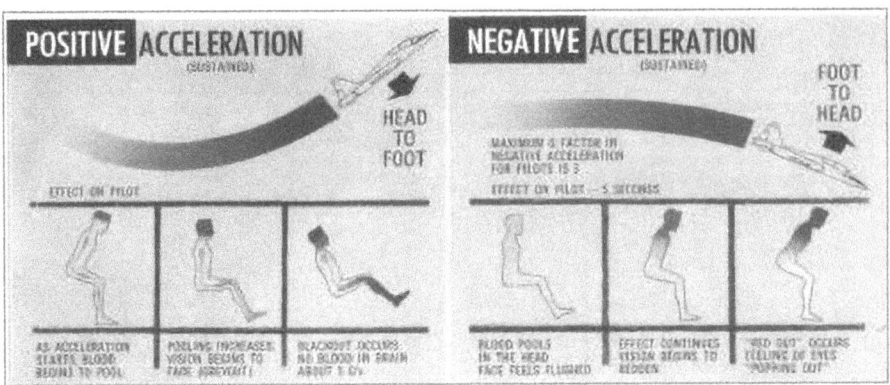

The Flight Surgeon's understanding of these biomechanical principles is fundamental to advising on aircraft maneuver limitations, developing protective strategies, and managing G-induced physiological responses.

Noise: Aircraft noise, both in flight and during ground operations, constitutes another significant hazard. Exposure to hazardous noise can lead to noise-induced hearing loss, excessive fatigue, and interference with voice communication. The effects of noise are categorized by thresholds: hearing, interference with rest/sleep, speech communication, hearing damage risk, and pain. The "Damage Risk Criterion" outlines sound pressure levels and frequencies that, if exceeded continuously, pose a risk of hearing impairment. Jet aircraft produce broadband "white noise" from engines and slipstream, while reciprocating engines generate discontinuous, low-frequency noise. Ground operations, particularly jet engine run-ups, expose ground crews to extremely high and hazardous noise levels (e.g., 140 dB near engines). Protection strategies involve reducing noise at the source and providing personal protective devices like earplugs (e.g., V-51R) and earmuffs. Flight Surgeons are responsible for implementing noise surveillance programs, issuing protective gear, and indoctrinating personnel on noise hazards and protection techniques. They also address subjective effects like fatigue, irritability, and somatic symptoms attributed to noise exposure.

Toxic Gases and Vapors: The presence of toxic gases and vapors in aircraft compartments poses a critical threat to aircrew safety and performance. Contamination can arise from exhaust gases, hydraulic fluid mist, fuel vapors, coolants, oil fumes, fire extinguishants, cargo, and thermal decomposition products of electrical insulation.

- **Carbon Monoxide** (CO) from piston engine exhaust is odorless and can cause hypoxia by binding with hemoglobin, interfering with oxygen transport. Symptoms include headache, vertigo, and impaired judgment, exacerbated at altitude. Flight Surgeons must educate aircrew to suspect CO when exhaust fumes are noted and to use 100% oxygen masks.

- **Oxides of Nitrogen** (e.g., NO2) from jet assist take-off units using nitric acid are highly irritating and can cause lung edema, with dangerous effects delayed hours after exposure.

- **Aviation Fuels** (gasoline and JP fuels) release toxic vapors that can cause narcotic effects, irritation, and central nervous system symptoms, particularly from tetraethyl lead in gasoline.

- **Hydraulic Fluid Vapors**, especially from castor oil-based fluids, contain toxic alcohols and glycol derivatives that can cause eye/respiratory irritation, headache, vertigo, and impaired judgment.
- **Coolant Fluid Vapors** (ethylene glycol) primarily cause respiratory irritation, though ingestion is more toxic.
- **Oil Fumes** from leaking hoses on hot engine parts produce irritants like aldehydes, causing symptoms similar to CO poisoning.
- **Fire Extinguishants** like carbon dioxide cause labored breathing and neurological symptoms at high concentrations, while chlorobromomethane (CB) acts as a narcotic and can be highly toxic upon thermal decomposition.
- **High Energy Fuels** (boron hydrides like pentaborane and decaborane) are extremely toxic central nervous system irritants, requiring stringent preventive measures and immediate decontamination upon exposure.

The Flight Surgeon must educate personnel on these toxins, symptoms of exposure, emergency first aid, and preventive measures like using detection devices and appropriate respiratory protection.

Addressing Prolonged Stress and Fatigue

Prolonged stress and fatigue represent critical challenges to aircrew effectiveness and safety in aerospace operations. The extended ranges, increased performance, and complexity of modern aircraft, coupled with long mission durations, often exceed the human physiological and psychological capacities. Fatigue contributes significantly to aviation accidents by leading to erroneous decisions and performance decrements.

What is Fatigue? Fatigue is defined as a detrimental alteration or decrease in skilled performance, directly related to the duration or repetitive nature of a task. It is aggravated by physical, physiological, and psychic stress. Physiologically, fatigue involves energy depletion and metabolite buildup, but for aircrew, it extends to a "central fatigue" affecting cognitive function.

Single Mission Skill Fatigue: This type of fatigue arises from the wearing repetition of tasks during long or multiple short missions. Symptoms include lassitude, disinclination for further activity, and disappear with adequate rest. Intrinsic factors like boredom, responsibility, frustration, anxiety, and even fear, alongside extrinsic factors such as hypoxia, temperature extremes, noise, vibration, G-forces, and restricted movement, contribute to skill fatigue.

Manifestations include the need for larger stimuli for appropriate responses, errors in timing, overlooking important task elements, loss of accuracy and smoothness in control, unawareness of accumulating errors, and an increase in control movements. Subjectively, individuals experience increased physical discomfort, irritability, and a projection of this irritability onto the aircraft. Flight Surgeons must identify these signs to prevent critical errors, such as reduced visual scanning or impaired judgment, which can lead to mid-air collisions or loss of control. Understanding concepts like the "Threshold of Indifference," "Anticipation Span," and the impact of "Speed and Load" in high-performance aircraft helps explain how fatigue degrades performance and makes smooth, safe flight more challenging.

Chronic Flying Fatigue: This is a cumulative phenomenon resulting from incomplete physical and mental recuperation between repeated missions. It can develop rapidly, often within one or two weeks of a demanding mission program. Studies from historical operations like the Berlin and Tokyo Airlifts highlighted common symptoms: tiredness, apprehension, increased drinking, weight loss, bickering, and numerous minor physical complaints. Commanders observed operational deficiencies such as bumpier landings, careless taxiing, clumsy handling of controls, and flight planning errors. Key contributing factors identified were lack or disorder of sleep, waiting between flights, poor living conditions, long hours, and unsatisfactory ground and air organization. Chronic fatigue is exacerbated by disrupted diurnal rhythms, extended duty times, and the monotony of routine operations compared to the stimulating challenge of emergency situations.

Management: Effective management of fatigue is primarily preventive, as there are no objective real-time indicators. This involves comprehensive pre-flight, in-flight, and post-flight control of the aircrew and their environment. Human factors engineering in aircraft design, improved cockpit ergonomics, and simplified instrument data presentation contribute significantly. Command leadership is vital in mission planning, balancing crew loads, and ensuring adequate support facilities to maintain morale and motivation. Flight Surgeons play a key role in educating aircrew on mature self-discipline, the importance of sleep, proper nutrition, regular physical conditioning, and managing personal problems. They also monitor compliance with crew rest policies and identify factors contributing to fatigue, such as post-alcoholic hangovers. During flight, measures like extra crew members for rotation and sleep, comfortable crew facilities, and appropriate in-flight nutrition help sustain performance. While stimulant drugs like dextro-amphetamine can temporarily postpone fatigue, they are not a substitute for rest and must be used with extreme caution under strict Flight Surgeon control due to potential side effects and judgment impairment. For future space operations, fatigue management will remain a major challenge due to prolonged missions, monotony, and minimal input workloads, underscoring the Flight Surgeon's role as a technical advisor in developing fatigue-preventing systems.

Nutritional Considerations for Aircrew

Nutrition is a fundamental pillar for the sustained effectiveness of all fighting forces, and its application is particularly vital within the USAF to meet the rigorous flight requirements for aircrews. The structured feeding procedures designed to achieve optimal nutritional goals are categorized into three distinct phases: the Ground Feeding Program, which caters to personnel stationed at Air Force Bases; the Flight Feeding Program, specifically tailored for airborne situations; and the Survival Feeding Program, designed to provide sustenance to airmen isolated in hostile or remote environments.

Ground Feeding: The ground feeding of Air Force personnel adheres to a standardized system common with the U.S. Army, where ration supplies are procured and distributed through Army quartermaster channels. Menus are planned months in advance by a Joint-Army-Air Force Master Menu Board, ensuring compliance with nutritional standards (AFR 160-95). Local adjustments to menus, coordinated by base menu boards including the Flight Surgeon, account for climatic, personnel, and supply conditions. The primary standard ration is Field Ration A, offering fresh, perishable foods when kitchen and refrigeration facilities are available. In their absence, the operational B Ration provides canned or dehydrated nonperishable items. Smaller units may use the 5-in-1 Ration, and specialized supplements are available for hospitals or combat conditions.

Flight Feeding: Flight feeding is categorized into pre-flight, in-flight, and post-flight phases, all extending the basic ground nutrition program. It has become increasingly necessary due to the extended ranges and performance of modern aircraft, which often disrupt normal sleeping, eating, and drinking habits. The primary goal is to assist aircrews and passengers in adjusting to these work demands, preventing "nonfeeding" which contributes to fatigue, human error, and potential accidents. Flight feeding properly "refuels" operators with nutrients, akin to aircraft refueling. The three phases are consecutive; ground-kitchen facilities implement pre-flight and post-flight feeding, while in-flight feeding faces restrictions on food preparation and consumption. Airmen are trained to follow good dietary patterns, and food servicing is often an individual responsibility.

- **Pre-flight Feeding:** This phase mandates consumption of a freshly prepared, balanced meal one to two hours before takeoff, regardless of time of day, to promote relaxation and aid digestion. Fighter pilots and bomber crews may require specific diet control to minimize gas pains and enhance high-altitude effectiveness. Meals high in carbohydrates and free of flatulence-producing foods (e.g., cabbage, dried peas, carbonated drinks) are recommended. Gum chewing is discouraged due to air swallowing. Fresh fruits and juices are permitted to prevent vitamin C depletion. High-fat, spiced, or poorly cooked foods are avoided. Alert-crew feeding, a special pre-flight situation, allows commanders to establish dedicated dining facilities, utilizing authorized food items or pre-cooked frozen meals.

- **In-flight Feeding:** A relatively new development, in-flight feeding became crucial during World War II with longer mission durations. It balances nutritional needs with limited aircraft space and equipment. There is no single method or standard food packaging; instead, simplicity, ease of support, and variety are key. Drinking fluids are supplied on flights over three hours (one quart per crew member per 16 flight hours), and flight lunch storage and heating facilities are scheduled for flights over six hours. "Flight duration" for feeding purposes is the total time from the last ground meal to the end of post-flight debriefing. Crew appetites often decrease during long flights due to work concentration, noise, vibration, and decreased oxygen, which can reduce digestive processes. Taste acceptability may also vary, with potatoes, vegetables, and salads rated lower in air. Monotony of diet is a problem for aircrews. Small, sugar-yielding food supplements are desirable between meals (recommended six hours apart), and beverages are crucial. Seven types of flight meals are authorized, with others requiring Headquarters USAF approval.

  o **Food Packet, Individual, In-Flight (IF):** Canned items for bases without fresh food or aircraft without storage. Ten different menus, each a complete meal (1,200 calories), with meat, fruit, bread, dessert, and accessories (coffee, tea, sugar, gum). Edible cold but enhanced by heating. Versatile, nonperishable, and requires minimal servicing equipment.

  o **Precooked Frozen Meal:** Main dishes are centrally procured, supplemented by fresh items from flight kitchens. Twelve menus (dinner and breakfast options) require aircraft ovens and refrigerators. Meals are dated and have a 9-month shelf life. A frozen water-filled vial indicates spoilage if melted and flowed along its axis.

  o **Sandwich Meal:** Most common, prepared by dining halls or flight kitchens. Fresh sandwiches (sliced meats, chicken, turkey, boneless), milk, juices, fresh fruit/desserts, and other items. Wrapped immediately and refrigerated below 40°F. No gravy, chopped egg, or chopped meat fillings due to bacterial food poisoning risk. Must be consumed within five hours of preparation or destroyed.

  o **Precooked Hot Meal and Breakfast Meal:** Rarely used in-flight. Hot meals are prepared on the ground, kept warm in insulated containers/ovens, but have poor keeping qualities. Breakfast meals are ready-to-eat items assembled in-flight.

  o **Bulk Issue for Preparation Aloft:** Food components are issued in bulk for other flight types or procured by aircraft commanders from commissaries.

  o **Bite-Size Meal:** For jet aircraft when other meals are impractical. Components must be bite-sized, suitable for hand-eating, and consumed within five hours. Includes beverages, cooked meats, fruit/candy, and optional items.

  o **Foil-Pack Meal:** Authorized for specific operations (e.g., radar picket patrol) in large aircraft with space and power. Uses hand-assembled, uncooked aluminum-foil packages (meat, two vegetables, hot roll, dessert), sealed and refrigerated until final cooking in specialized ovens. Highly acceptable and popular.

- **Beverages:** Essential for preventing dehydration, which lowers efficiency, especially in hot climates or high altitudes. Cool water, coffee, tea, chocolate milk, and juices are popular. Drinking fluids are supplied in all aircraft flying over three hours, with one quart per person per 16 flight hours. Insulated jugs (e.g., CNU-2/C) keep liquids hot or cold, operating on DC or AC power, or with wet ice. Special crew position water bottle assemblies (one-quart horizontal, two-quart vertical) are standard for fixed positions. A can-piercing drinking device allows direct consumption from commercial cans.

- **Feeding Procedures and Equipment:** Developmental studies focus on fighter in-flight feeding (reduced caloric intake, liquids via special dispensers), food servicing equipment, and microbiology. The B-4 in-flight oven heats frozen/foil-pack meals and canned items, with six independently heated shelves and a warming element. The B-3 oven warms canned IF components and soups. Galleys, integrated frameworks for storage and work, are designed for each aircraft type, containing rectangular liquid containers, ovens, hot cup brackets, hot cups, dispensers, refuse containers, can openers, and refrigerators. Hot cups (28V DC or 115V AC) heat water for beverages and can be used to warm liquid/semi-solid foods. Mechanical refrigerators (SR-4, SR-6, SR-6A, SR-10) maintain specific temperature ranges for fresh and frozen foods. Dry ice refrigerators (Type B-1) hold frozen meals for extended periods. Minor items include disposable pasteboard trays and accessory packets (plastic utensils, salt, pepper, napkins).

- **Microbiology of Flight Meals:** Food-borne infections are a serious concern, especially in-flight. Continuous preventive control involves ground kitchen sanitation, refrigerated storage, and adherence to time-temperature factors for bacterial growth. The "danger zone" for bacterial multiplication is 50-130°F, with a minimum incubation period of five hours. Sanitary practices, rapid cooling or freezing, and heating above 130°F are critical. While aircraft food heating equipment often destroys bacteria, enterotoxins may remain active. Individual packaging and sanitized containers offer supplementary protection.

- **Post-flight Feeding:** This phase is crucial for stimulating physiological recovery and morale, shortening mission turnaround times, and preventing chronic fatigue. Convenient flight-line kitchen facilities are requisite. Light refreshments may precede a more complete protein-predominant dinner meal, aiding in tension reduction after long hours of concentration.

Flying Safety and Accident Investigation

The medical aspects of flying safety impose many critical requirements, directly impacting the overall aircrew effectiveness program. The Flight Surgeon plays an indispensable role in ensuring a robust and responsive safety framework, primarily through strategic planning, comprehensive training, and meticulous investigation of incidents.

Problems related to emergency crash procedures and casualty management must be met with proper planning and rigorous training. This involves not only equipping personnel with the necessary skills to respond to a crash but also ensuring that medical protocols are in place to manage casualties effectively and expeditiously. A fundamental consideration within any flight safety research program is the thorough investigation of aircraft crashes. The Flight Surgeon's involvement in such investigations frequently predominates, primarily because of the high incidence of human factors contributing to the causation of accidents. Human factors encompass a wide array of issues, including physiological limitations, psychological states, decision-making errors, and interactions with aircraft systems. Given this complex interplay, the Flight Surgeon's expert analysis is critical in identifying root causes and recommending preventative measures. It is therefore imperative that the flying safety program of every organization receives the active and unwavering support of the Flight Surgeon, whose insights are vital for continuous improvement and accident prevention.

Beyond immediate crash response, the Flight Surgeon's responsibilities extend to the assessment of injuries sustained during emergency escape procedures. For instance, in parachute jumps, while relatively few injuries or fatalities are sustained by trained parachutists, poor landing technique accounts for a significant portion of nonfatal injuries in emergency bailouts. Thus, familiarization with bailout procedures and landing techniques is crucial. The wearing of appropriate footgear, such as jump-type boots, is advisable to prevent foot and ankle injuries, especially given that low-quarter shoes are often lost during windblast or parachute opening shock and provide inadequate ankle support. The Flight Surgeon contributes to this by advising on protective gear and training protocols to minimize such injuries.

Moreover, the Flight Surgeon must be aware of the hazards associated with injuries from ejection seats. Many injuries sustained during ejections are compression fractures of vertebrae, often involving the anterior vertebral lips, indicative of spinal flexion at the moment of ejection. Proper body positioning—buttocks pressed against the seat back, lumbar spine straightened, and head pressed against the headrest with chin tucked in—is critical to provide a solid structure for transmitting ejection forces and minimizing injury. Poor positioning, exacerbated by aircraft attitude, lack of preparation time, or attempts to reach controls, significantly increases injury risk. The Flight Surgeon's role includes training pilots on correct ejection posture and ensuring proper use of cushions, which, if too soft or compressible, can increase peak G-forces on the occupant.

The Flight Surgeon also has a vital role in understanding crash landings and accidents. While it is often believed that persons involved in aircraft accidents are either fatally injured or completely uninjured, studies show a significant percentage suffer major non-fatal injuries, particularly vertebral fractures. These injuries are more frequent in jet aircraft, often due to vertical decelerative forces ("slap down" in front-seat occupants) and direct traumatic injury from aircraft structural collapse. The most commonly affected vertebrae are at the thoracolumbar junction (T-12, L-1, L-2). The responsible medical officer must be prepared to recognize and manage these injuries, emphasizing careful removal of pilots from cockpits to prevent spinal cord transection or severe injury. Devices like sling harnesses, which permit removal of the pilot in a seated posture, are crucial for minimizing further spinal trauma. These considerations underscore the Flight Surgeon's multifaceted role in accident investigation and safety enhancement, from pre-flight preparation to post-crash management.

Aircraft and Equipment Design Considerations for Flight Surgeons

The Flight Surgeon's purview extends critically to the design and operational characteristics of aircraft and their associated equipment, all in the overarching interest of combat effectiveness and aircrew well-being. This involves a proactive and analytical approach to identify potential issues and contribute to their resolution.

The recognition and evaluation of problems related to aircraft and equipment design necessitate close observation of the flyer within their specific crew position. This hands-on, observational approach allows the Flight Surgeon to understand the practical implications of design choices on human performance and safety. The Flight Surgeon is inherently concerned with any aspect of aircraft or equipment design that impacts safety, comfort, well-being, and efficiency. This concern is particularly pronounced from the standpoints of psychological, physiological, and anatomical considerations, as these directly influence a crew member's ability to operate effectively for extended periods under demanding conditions.

In studying these complex problems, the Flight Surgeon may become aware of potential solutions or "fixes." These could range from minor ergonomic adjustments to significant design modifications. Furthermore, the Flight Surgeon may invoke the important administrative procedure of the "Unsatisfactory Report." This formal reporting mechanism is crucial for documenting equipment deficiencies or design flaws that negatively affect aircrew effectiveness. Perhaps even more importantly, the Flight Surgeon is uniquely positioned to report special observations gleaned from operational experience. When these observations are properly forwarded to relevant research and development agencies, they can result in significant improvements and important developments in future aircraft and equipment.

For example, in addressing the impact of accelerative forces, Flight Surgeons contribute to the design of protective devices. The anti-G suit, which applies external counter-pressure to prevent blood pooling and increase arterial pressure during positive G maneuvers, is a direct result of aeromedical research translated into design. Similarly, the evolution of flying helmets, from rudimentary leather designs to rigid plastic shells with sophisticated communication and visor systems (e.g., P-series and HGU helmets), reflects the continuous effort to provide head protection against impact, windblast, and enhance communication, comfort, and oxygen mask retention in high-speed flight. The Flight Surgeon's input ensures that these devices are not only effective but also comfortable and compatible with other personal equipment.

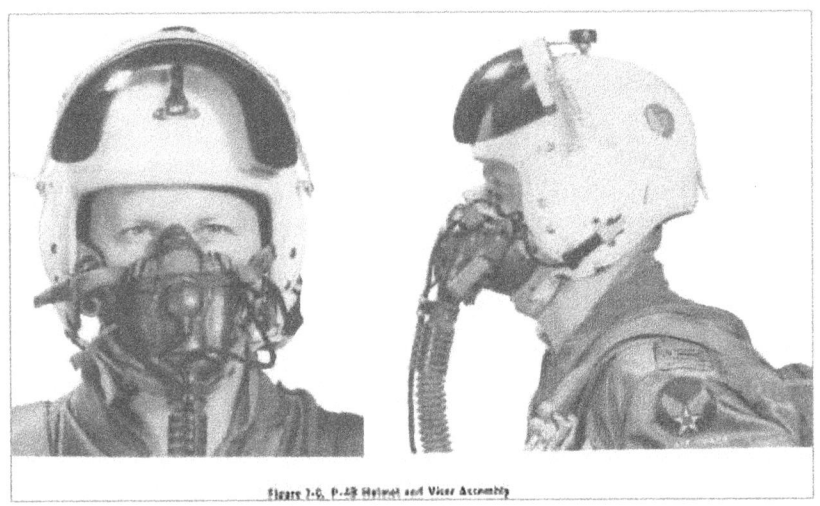

Figure 1-6. P-4B Helmet and Visor Assembly

In the context of emergency escape from aircraft, the Flight Surgeon's insights are indispensable for the design of ejection seats and parachute systems. Early observations on the incapacitating effects of G-forces during escape led to the development of ejection seats, which continue to evolve with rocket catapults to improve low-altitude escape performance. The design of these systems must consider human tolerance to vertical acceleration, rate of onset, and duration of forces, ensuring optimal body positioning (e.g., straightened spine) to prevent vertebral fractures. Furthermore, advancements in parachute technology, such as quarter bags to reduce opening shock and automatic release devices (F-1, F-1A, F-1B) with barometric and time-delay settings, directly integrate aeromedical principles to enhance survivability during high-altitude bailouts and free-falls. The development of escape capsules for very high-speed and high-altitude aircraft also incorporates Flight Surgeon recommendations to mitigate hazards like windblast, cold, and low pressure.

The Flight Surgeon also influences equipment design related to nutrition. The development of in-flight feeding equipment, such as insulated jugs for beverages, crew position water bottle assemblies, and can-piercing drinking devices, ensures that aircrew can maintain hydration and caloric intake even in cramped and demanding cockpits. The design of specialized ovens (e.g., B-4 and B-3 ovens) and galleys for heating and storing various types of meals (IF packets, precooked frozen, foil-pack) demonstrates a direct link between nutritional requirements and aircraft interior design. These considerations aim to overcome the practical limitations of aircraft space and power while ensuring food safety and palatability.

These various categories of problems concerning the flyer, the recognition of these problems, and the accomplishment of activities that solve them constitute the core of the aircrew effectiveness program. This program is unequivocally the backbone of military aviation medicine, emphasizing the Flight Surgeon's role not just as a clinician but as a vital contributor to aircraft and equipment development. Much of the specific knowledge essential for the accomplishment of such a comprehensive program is detailed in the subsequent chapters of this manual, further underscoring the depth and breadth of the Flight Surgeon's impact on optimizing aircrew performance and well-being.

References

Armstrong, H. G. (2017). Aircrew Maintenance. In "Aerospace Medicine". Williams and Wilkins Co.

Graybiel, A. (2020). Evolving Challenges for Pilots in Aviation Medicine. "Journal of Aviation Medicine", 27, 397–406.

Ogle, D. C. (2022). Strategic Management of Aviation Medical Services. "USAF Medical Service Digest", 6, 14–20.

Stratton, K. L. (2019). Medical Management of Airline Pilots. "Journal of Aviation Medicine", 25, 630–636.

USAF. (2023). "Aircrew Effectiveness Program (AFR 160-69)". U.S. Air Force.

Chapter 2: Physiological Impact of Hypoxia and Oxygen Deprivation

Respiration, fundamentally, represents the interaction between an organism and the gaseous environment. In humans, this vital process is categorized into two primary phases: the pulmonary phase and the tissue phase. The pulmonary phase encompasses the exchange of gases between the ambient atmosphere and the alveolar air, as well as between alveolar air and the blood within the pulmonary capillaries. The subsequent tissue phase involves the critical exchange of gases between the body's cells and the blood in the tissue capillaries. A comprehensive understanding of these phases is crucial for comprehending the physiological effects of decreased partial pressure of oxygen, particularly in aerospace operations.

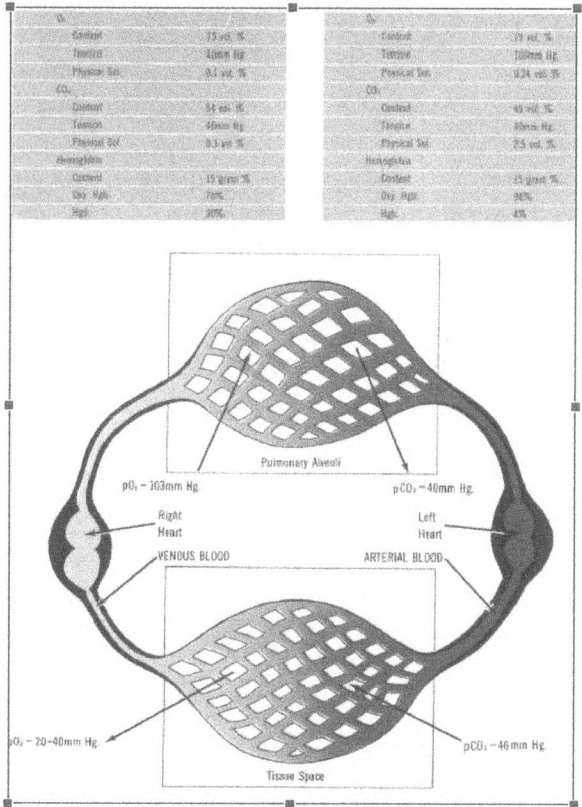

Respiratory Physiology Overview

The respiratory system facilitates the continuous exchange of gases, ensuring adequate oxygen supply to the body and efficient removal of carbon dioxide. A key aspect of this system is the lung volumes and capacities, which are dynamic measurements reflecting the mechanics of breathing. The total volume of air within the lungs, known as total lung capacity, is further subdivided into distinct primary lung volumes and lung capacities. These subdivisions are instrumental in studying pulmonary function under various conditions, including health, disease, and the abnormal environmental pressures encountered in flight. The end of a quiet expiration serves as the standard reference point for quantitatively measuring these subdivisions.

At the termination of a quiet expiration, the elastic recoil force exerted by the lung is approximately balanced by the expansile tendency of the chest wall. This state is often referred to as the equilibrium point. There are four primary lung volumes: 1. **Tidal Volume (TV):** The volume of air exchanged during a single breath. In a resting state, this averages approximately 500 cubic centimeters (cc). 2. **Inspiratory Reserve Volume (IRV):** The maximum additional volume of air that can be inspired following a normal, resting inspiration. This volume is highly variable and depends on the tidal volume. 3. **Expiratory Reserve Volume (ERV):** The maximum volume of air that can be forcibly expired after a normal expiration. The average value for ERV is about 1200 cc. 4. **Residual Volume (RV):** The amount of air that remains within the lungs even after a maximal expiratory effort. This typically averages around 1200 cc, representing 20% to 25% of the total lung capacity.

Combinations of these primary lung volumes form lung capacities, which often provide a more accurate reflection of the functional compartments of the lung. The four major lung capacities are: 1. **Total Lung Capacity (TLC):** The sum of all four primary lung volumes, averaging about 6000 cc. 2. **Inspiratory Capacity (IC):** The maximum volume of air that can be inhaled from the end of a quiet expiration. It is the sum of the tidal volume and the inspiratory reserve volume, averaging approximately 3600 cc. 3. **Vital Capacity (VC):** The maximum volume of air that can be exhaled following a maximal inspiration. This typically averages around 4800 cc and represents the sum of the inspiratory reserve volume, tidal volume, and expiratory reserve volume. 4. **Functional Residual Capacity (FRC):** The amount of air remaining in the lungs following a normal tidal expiration.

These lung volumes and capacities can be measured using a spirometer or similar calibrated recording device. It is important to note that the average values provided are approximations, as individual values are influenced by factors such as age, sex, height, and weight. More precise individual estimations can be achieved through regression formulas that account for these variables, often expressed as a "Percent of predicted normal."

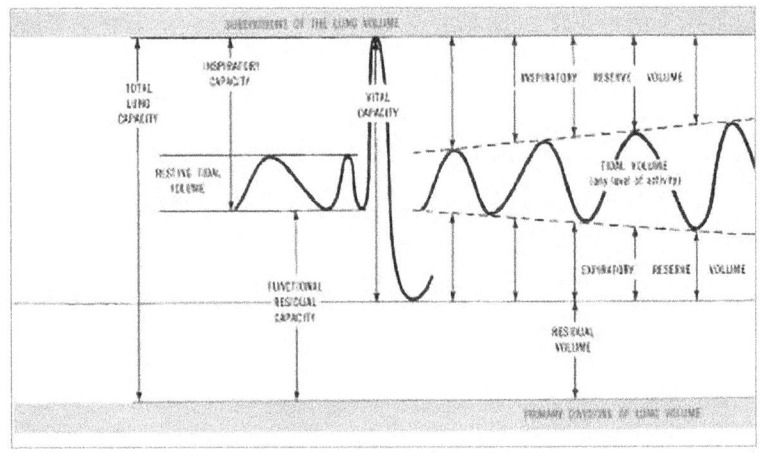

Gas Exchange Mechanisms

The major physiological functions of the lungs are categorized into ventilation, diffusion, and perfusion, all critical for effective gas exchange.

Ventilation refers to the mass movement of air into and out of the lungs, or the process by which alveolar air is continuously diluted by atmospheric air. Adequate ventilation relies on the creation of a pressure gradient between the alveoli and the external atmosphere, generated by the bellows-like action of the chest wall and diaphragm. The patency of the airways, the functional integrity of the respiratory center in the medulla, and the strength of the intercostal, abdominal muscles, and diaphragm are all vital for maintaining effective ventilation. Furthermore, an even distribution of inspired gases throughout the lung is of paramount importance. Impairments such as secretions, bronchial narrowing, or occluding masses can lead to uneven alveolar ventilation, where some areas are hyperventilated while others are underventilated. This uneven distribution can cause the lung to function as a collection of compartments, each ventilating at its own rate. For example, roughly 50% of normal individuals exhibit relatively slow ventilation in a lung compartment accounting for 10% to 50% of their functional residual capacity. In severe obstructive emphysema, up to two-thirds of the functional residual capacity might receive only 10% of the total ventilation.

Diffusion describes the mechanism by which respiratory gases traverse the alveolar-capillary wall, moving between alveolar air and the pulmonary capillary blood. Carbon dioxide diffuses significantly faster than oxygen, approximately 25 times more rapidly. However, the pressure gradient for oxygen across the alveolar wall is about 25 times greater than that for carbon dioxide. Conditions such as fibrosis, granuloma, edema, or exudate within the alveoli or the alveolar-capillary wall can severely impede diffusion, potentially leading to hypoxia or carbon dioxide retention, or both. Specific diffusion abnormalities, such as granulomatous involvement of the alveolar wall in pulmonary sarcoidosis, are termed "alveolar-capillary block syndromes." In diseases that cause diffusion abnormalities, oxygen diffusion is generally compromised earlier and to a greater extent than carbon dioxide diffusion. This is due to the higher diffusibility of CO_2 and the ability of normally functioning lung areas to compensate for CO_2 retention in diseased regions by increasing minute ventilation. The relatively flat nature of the oxyhemoglobin dissociation curve in the physiological range also means that large increases in ventilation have minimal impact on oxygen saturation.

Perfusion refers to the flow of blood through the lung capillaries. This process is not always uniform, even in healthy individuals, and can vary greatly in diseased states. Uneven perfusion, especially when combined with uneven ventilation, can have serious consequences. Areas of the lung that are well-ventilated but poorly perfused effectively increase dead space, contributing minimally to gas exchange. Conversely, areas that are well-perfused but poorly ventilated function as right-to-left vascular shunts, as blood flowing through them retains its venous characteristics. Minor disturbances in ventilation-perfusion relationships can occur during flight, particularly when accelerative "G" forces redistribute blood flow within the lungs. For instance, footward G forces lead to engorgement of the lower lung lobes, while headward G forces engorge apical regions.

Composition of Atmospheric and Alveolar Air

The composition of atmospheric air is fundamental to understanding respiratory physiology at altitude. Dry atmospheric air, by volume, consists of 20.94% oxygen, 79.03% nitrogen, and 0.03% carbon dioxide. Trace amounts of rare gases are present alongside nitrogen but hold no physiological significance. Crucially, the relative percentage composition of dry atmospheric air remains largely consistent up to 70,000 feet, without significant variations across latitudes.

However, expressing gas quantities in percentages at various altitudes is insufficient because it does not reflect the molecular concentration, which dictates gas availability to the body. Molecular concentration is best represented by partial pressure. According to Dalton's Law, the partial pressure of a gas in a mixture of non-interacting gases is equal to the pressure it would exert if it alone occupied the same volume. The total pressure of a gas mixture is the sum of the partial pressures of its individual components. For moist air, this is expressed as:

$B = pO_2 + pN_2 + pCO_2 + pH_2O$ where B is the total barometric pressure, and pO_2, pN_2, pCO_2, and pH_2O are the partial pressures of oxygen, nitrogen, carbon dioxide, and water, respectively.

At sea level, the total standard barometric pressure is 760 mm Hg (14.7 psi). For dry air, the partial pressure of oxygen at sea level is calculated as: (20.94 / 100)

760 { mm Hg} = 159 { mm Hg (3.1 psi)} The partial pressure of nitrogen is: (79.03 / 100) 760 { mm Hg} = 601 { mm Hg (11.6 psi)} Partial pressures of other gases can be calculated similarly.

Upon inhalation, atmospheric air passes through the nasal passages into the trachea, becoming saturated with water vapor. This humidified air then mixes with the existing alveolar air. Within the alveoli, a continuous gaseous interchange occurs across an interface: newly inspired air yields oxygen and receives carbon dioxide, while existing alveolar air receives oxygen and yields carbon dioxide. Consequently, expired air contains less oxygen and more carbon dioxide than inspired air. Expired air is not a true representation of alveolar conditions, as it is a mixture from both alveoli and anatomical dead space. The partial pressure of oxygen in the alveoli is the critical determinant of how much oxygen reaches the blood.

Alveolar partial pressures for breathing air at sea level are approximately: *

$pO_2 = 103$ { mm Hg} * $pCO_2 = 40$ { mm Hg} * $pH_2O = 47$ { mm Hg} * $pN_2 = 570$ { mm Hg}

Altitude (Feet)	Pressure mm Hg	Pressure p.s.i.	Altitude (Feet)	Pressure mm Hg	Pressure p.s.i.	Altitude (Feet)	Pressure mm Hg	Pressure lb/ft²
0	760.0	14.70	32500	201.0	3.89	66000	40.6	113.2
500	746.4	14.43	33000	196.3	3.80	68000	36.9	102.9
1000	732.9	14.17	33500	191.8	3.71	70000	33.6	93.52
1500	719.7	13.92	34000	187.3	3.62	72000	30.4	85.01
2000	706.6	13.66	34500	183.0	3.54	74000	27.7	77.26
2500	693.8	13.42	35000	178.7	3.46	76000	25.2	70.22
3000	681.1	13.17	35500	174.4	3.37	78000	22.9	63.8
3500	668.6	12.93	36000	170.3	3.29	80000	20.8	58.01
4000	656.3	12.69	36500	166.3	3.22	82000	18.9	52.72
4500	644.2	12.46	37000	162.4	3.14	84000	17.2	47.91
5000	632.3	12.23	37500	158.6	3.07			
5500	620.6	12.00	38000	154.8	2.99	86000	15.6	43.55
6000	609.0	11.78	38500	151.2	2.92	88000	14.2	39.59
6500	597.6	11.55	39000	147.5	2.85	90000	12.9	35.95
7000	586.4	11.34	39500	144.1	2.79	92000	11.7	32.7
7500	575.3	11.12	40000	140.7	2.72	94000	10.7	29.71
8000	564.4	10.91	40500	137.4	2.66			
8500	553.7	10.71	41000	134.1	2.59	96000	9.7	27.02
9000	543.2	10.50	41500	131.0	2.53	98000	8.8	24.55
9500	532.8	10.30	42000	127.9	2.47	100000	8.0	22.31
10000	522.6	10.11	42500	124.9	2.42	110000	5.0	13.92
10500	512.5	9.91	43000	121.9	2.36	120000	3.24	9.026
11000	502.6	9.72	43500	119.0	2.30			
11500	492.8	9.53	44000	116.2	2.25	130000	2.18	6.071
12000	483.3	9.35	44500	113.5	2.19	140000	1.51	4.213
12500	473.8	9.16	45000	110.9	2.14	150000	1.08	3.033
13000	464.5	8.98	45500	108.2	2.09	160000	0.787	2.190
13500	455.4	8.81	46000	105.6	2.04	170000	0.583	1.624
14000	446.4	8.63	46500	103.1	1.99	180000	0.433	1.206
14500	437.5	8.46	47000	100.7	1.95			
15000	428.8	8.29	47500	98.3	1.90			
15500	420.2	8.13	48000	96.0	1.86			
16000	411.8	7.96	48500	93.7	1.81			
16500	403.5	7.80	49000	91.5	1.77			
17000	395.3	7.64	49500	89.4	1.73			

				Pressure	
Altitude (Feet)	Microns Hg	lb/ft²
190000	321.6	0.8956
200000	238.6	0.6645
210000	174.9	0.4860
220000	125.9	0.3604
230000	88.69	0.2470
240000	61.02	0.1699
250000	40.9	0.1139
260000	26.55	0.07422

17500	387.3	7.49	50000	87.3	1.69
18000	379.4	7.34	50500	85.2	1.65
18500	371.7	7.19	51000	83.2	1.61
19000	364.0	7.04	51500	81.2	1.57
19500	356.5	6.89	52000	79.3	1.53
20000	349.1	6.75	52500	77.4	1.50
20500	341.8	6.61	53000	75.6	1.46
21000	334.6	6.47	53500	73.8	1.43
21500	327.6	6.33	54000	72.1	1.39
22000	320.8	6.20	54500	70.4	1.36
22500	314.0	6.07	55000	68.8	1.33
23000	307.4	5.94	55500	67.1	1.30
23500	300.8	5.82	56000	65.5	1.27
24000	294.4	5.70	56500	64.0	1.24
24500	288.0	5.57	57000	62.4	1.21
25000	281.8	5.45	57500	61.0	1.18
25500	275.8	5.33	58000	59.5	1.15
26000	269.8	5.22	58500	58.1	1.12
26500	263.8	5.10	59000	56.8	1.10
27000	258.0	4.99	59500	55.4	1.07
27500	252.4	4.88	60000	54.1	1.05
28000	246.8	4.77	60500	52.8	1.02
28500	241.4	4.67	61000	51.6	0.998
29000	236.0	4.56	61500	50.4	0.975
29500	230.6	4.46	62000	49.2	0.951
30000	225.6	4.36	62500	48.0	0.928
30500	220.4	4.26	63000	46.9	0.907
31000	215.4	4.17	63500	45.8	0.886
31500	210.4	4.07	64000	44.7	0.864
32000	205.6	3.98	64500	43.6	0.843

Day Pressure Only Gravity Constant

Altitude (Feet)	Microns Hg	lb/ft²
270000	17.34	0.04829
280000	11.49	0.03200
290000	7.843	0.02184
300000	5.493	0.01530

Sources of Data:

0—80,000 Brombacher Tables
NACA Report No. 538, 1935

80,000—300,000 NACA Tech Note
No. 1200, January 1947

Conversion Factors:

1 mm Hg = 0.019339 p.s.i.
1 p.s.i. = 51.715 mm Hg

At sea level, the combined pressure of pCO_2 and pH_2O is 87 mm Hg, occupying 11% of the lung volume. Oxygen occupies 14% (103/760), and nitrogen 75%. As altitude increases, the barometric pressure decreases, significantly impacting these partial pressures. At 18,000 feet, pCO_2 is 31 mm Hg, pH_2O remains 47 mm Hg, and the barometric pressure is 380 mm Hg. At this altitude, carbon dioxide and water vapor occupy (31+47)/380 , or 21% of the lung volume. At 38,000 feet, if pCO_2 reduces to 30 mm Hg (representing hypocapnia), carbon dioxide and water vapor would occupy (30+47)/155 , or 50% of the lung volume. This demonstrates that as ascent continues, the fixed partial pressure of water vapor and the metabolically determined partial pressure of carbon dioxide increasingly occupy a larger percentage of the total alveolar gas volume, thereby reducing the available volume for oxygen.

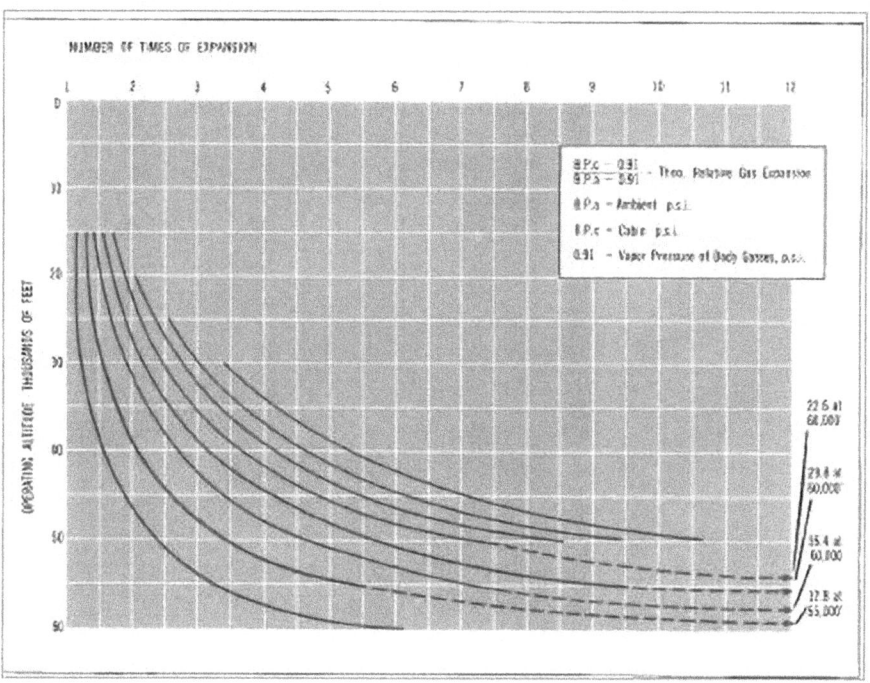

When breathing pure oxygen at 33,700 feet, the alveolar pO_2 is comparable to sea-level values while breathing air. However, above 34,000 feet, the alveolar

pO_2 begins to fall below sea-level values even with 100% oxygen. Beyond 40,000 feet, alveolar pO_2 decreases rapidly, falling below levels necessary for physiological safety. The critical altitude is 50,000 feet, where the ambient pressure of 87 mm Hg, even with 100% oxygen, provides no further contribution to respiratory needs. At 63,000 feet, the ambient pressure of 47 mm Hg equals the water vapor pressure in the body at normal body temperature, leading to tissue fluid boiling in an unprotected individual.

To counteract these effects, conventional pressure-demand oxygen systems are typically effective up to 43,000 feet and for emergencies up to 50,000 feet for short durations. Above 40,000 feet, these systems deliver 100% oxygen under increasing pressure. At extreme altitudes, the positive pressure required for safe physiological function surpasses lung tolerance, necessitating equivalent external counter-pressure, often provided by pressure suits. Cabin pressurization in aircraft aims to maintain a cabin altitude below 40,000 feet, eliminating the need for positive pressure breathing. For instance, an aircraft with a 6.55 psi pressure differential can fly at an ambient altitude of 63,000 feet while maintaining a cabin altitude of approximately 20,000 feet. The primary drawback of cabin pressurization is the risk of rapid decompression at altitude. In such a scenario (e.g., from a cabin altitude of 20,000 ft to 63,000 ft), a pressure suit providing adequate lung and counter-pressures becomes essential.

Number of Experiments	Type of Pressure Chamber	Volume of Container (cu. ft.)	Size of Opening (Square Inches)	PSI Differential Pressure	Simulated Altitude (ft.) Cabin	Simulated Altitude (ft.) Flight	Time of Decompression (sec.)	Rate of Decompression (psi per sec.)	Expansion of Body Gases (no. of times)	Rate of Expansion (vol. per sec.)	Comparable Size of Hole in 1000 cu. ft. Cabin	Comparable Size of Hole in 45 cu. ft. Cabin	Ascent in feet per sec.
Armstrong 26	1-Man Chamber	25	27	7.00	0	16,500	0.1	70	1.8	18			150,000
Dill 5	Chamber	..	..	11.9	0	40,000	365	.072	7.6	.046			242
J. J. Smith 5	1-Man Chamber	25	..	7.25	10,000	40,000	1.6	4.6	6.0	3.0			20,600
Sweeney 10	Pressure Suit	3	4	2.75	27,500	45,000	0.016	125	3.2	213	10	18	1,166,666
150	P-38 Mock-up	45	12	6.55	10,200	35,000	0.075	87	3.5	47	66	12	391,666
15	P-38 Mock-up	45	12	7.5	8,000	35,000	0.03	83	3.9	43	66	12	300,000
9	P-38 Mock-up	45	27	1.5	34,000	45,000	0.038	143	2.3	283		27	1,375,000
3	P-38 Mock-up	45	27	1.25	37,000	48,000	0.006	167	2.3	383		27	1,833,333
2	P-38 Mock-up	45	27	1.00	40,000	50,000	0.005	200	2.3	460		27	2,000,000

Oxygen transport within the blood relies significantly on mechanisms beyond simple solution. While only 0.24 cc of oxygen and 2.5 cc of carbon dioxide can dissolve in 100 cc of blood at sea level, the blood actually transports 18-20 cc of oxygen and 40-50 cc of carbon dioxide. This greatly enhanced capacity is primarily due to hemoglobin in red blood cells for oxygen, and bicarbonate ions in plasma and red blood cells for carbon dioxide. Oxygen reversibly combines with hemoglobin to form oxyhemoglobin, a process highly sensitive to the partial pressure of oxygen in the surrounding medium. This directly impacts oxygen delivery to tissues at various altitudes. A significant release of oxygen from hemoglobin occurs only when pO_2 falls below 40 mm Hg. The oxygen-carrying capacity of hemoglobin is also sensitive to blood pH (the Bohr effect); lower pH (due to higher pCO_2 in venous blood) facilitates oxygen release to tissues, while higher pH (due to lower pCO_2 in arterial blood) promotes oxygen uptake.

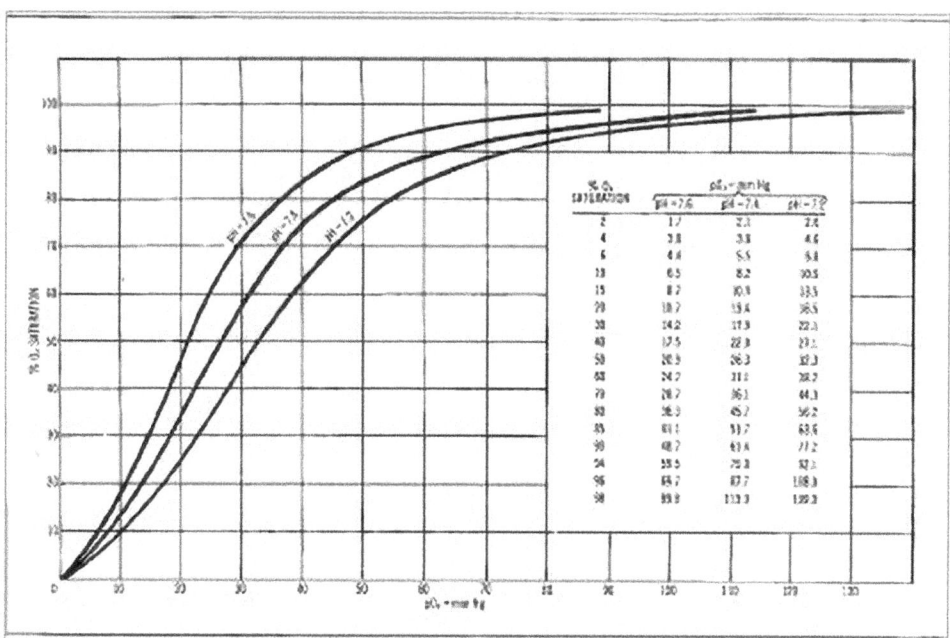

Regulation of Breathing

The control of respiratory rate and depth is a complex physiological process influenced by both chemical and psychic stimuli. In a resting individual, breathing typically occurs 12 to 16 times per minute without conscious effort, resulting in a minute respiratory volume of 6 to 8 liters. During physical exertion, the body's increased demand for oxygen and the need to eliminate carbon dioxide trigger a responsive increase in both ventilation and cardiac output. For a resting subject, minute respiratory volume remains relatively constant up to 30,000 feet. However, any increase in workload invariably leads to a rise in ventilation. A notable increase in minute respiratory volume is observed between 30,000 and 40,000 feet, even when breathing 100% oxygen. At 40,000 feet, the barometric pressure is so low that alveolar oxygen tension is equivalent to that at 10,000 feet while breathing ambient air, highlighting the physiological stress of altitude. Studies on bomber crews and fighter pilots have recorded resting minute respiratory volumes between 7 and 15 liters per minute (average 12 L/ min), increasing to 12-60 liters per minute (average 25 L/min) during active flight.

GROUND LEVEL
20,000 ft.
40,000 ft.
REST (SITTING)
NO LOAD
1000
2000
3000
4000
5000
6000
10
20
30
40
50
60
70

60,000 ft.
50,000 ft.
40,000 ft.
35,000 ft.
30,000 ft.
25,000 ft.
BAROMETRIC PRESSURE mm. Hg.
40
60
80
100
120
140
160
180
200
220
240
260
280
300
20
40
60
80
100
120
140
A
B
C
TIME OF CONSCIOUSNESS (SECONDS)
A. BENZINGER - 100% Oxygen at ambient pressure (rapid decompression from 40,000 ft.)
B. HALL - Air breathing (mask removal at altitude)
C. WILSON and COMFORT - Air breathing (rapid decompression from 10,000 ft.)

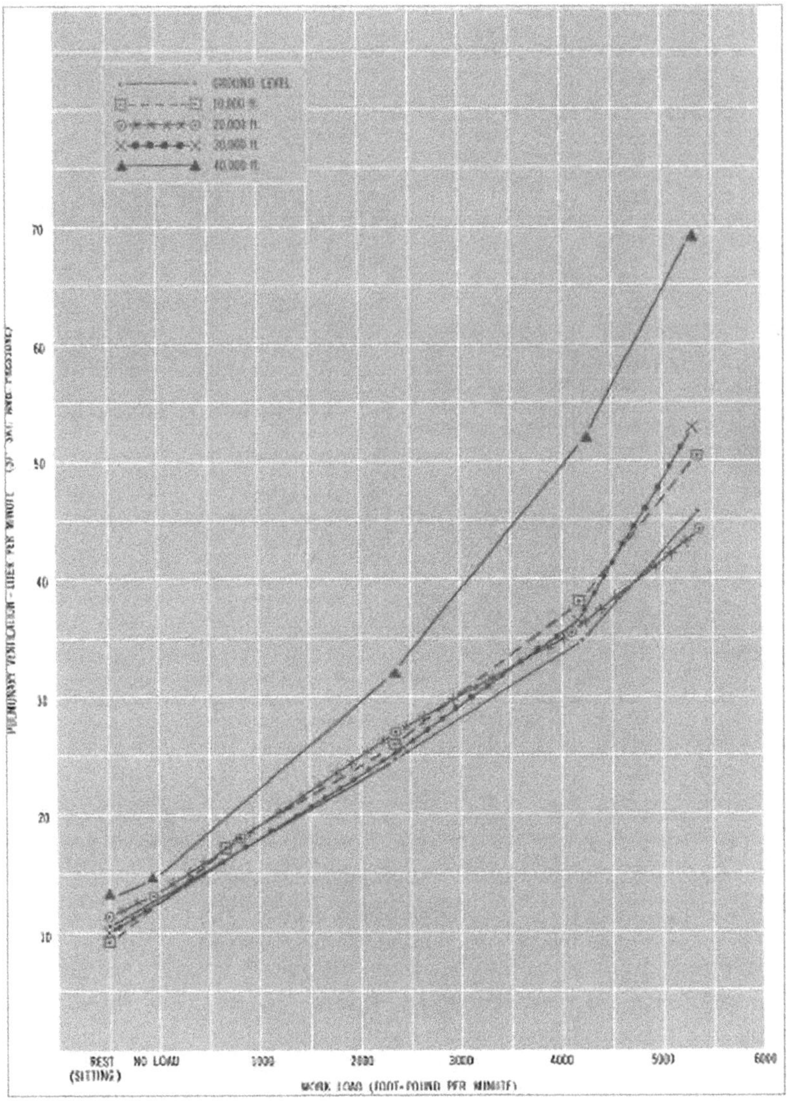

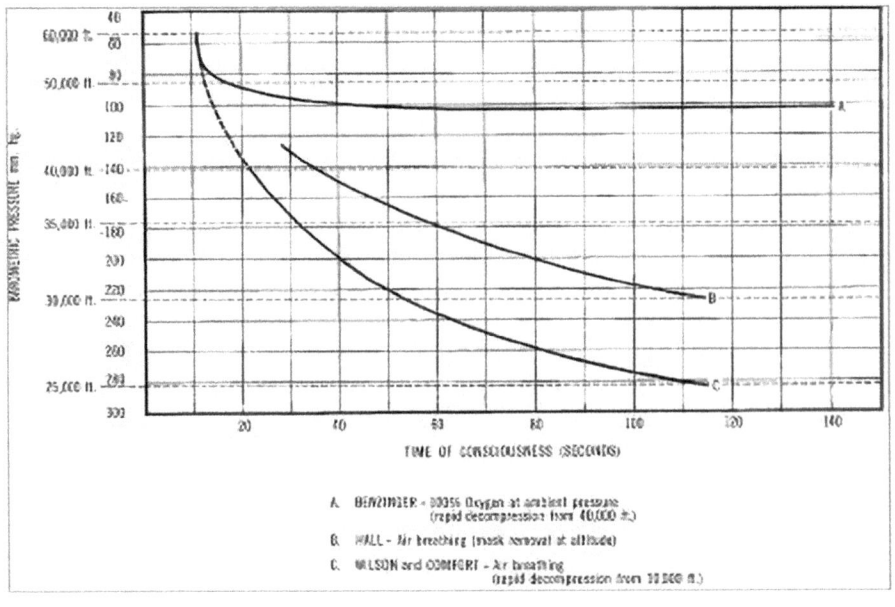

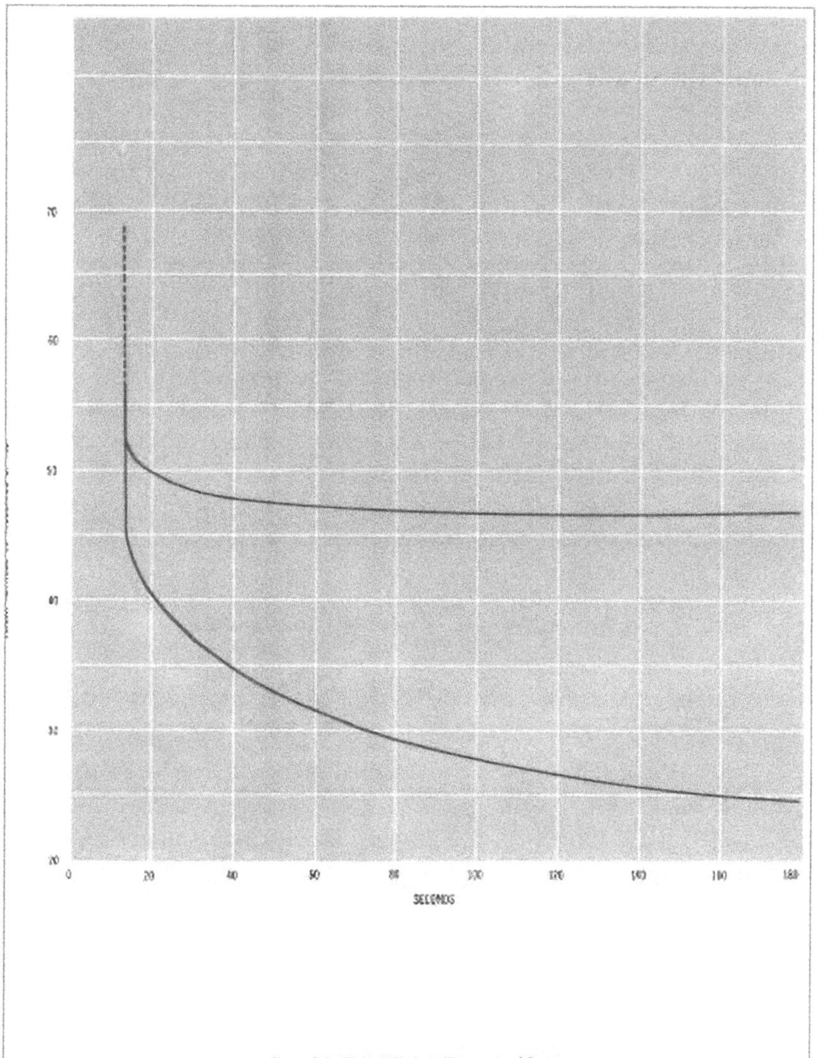

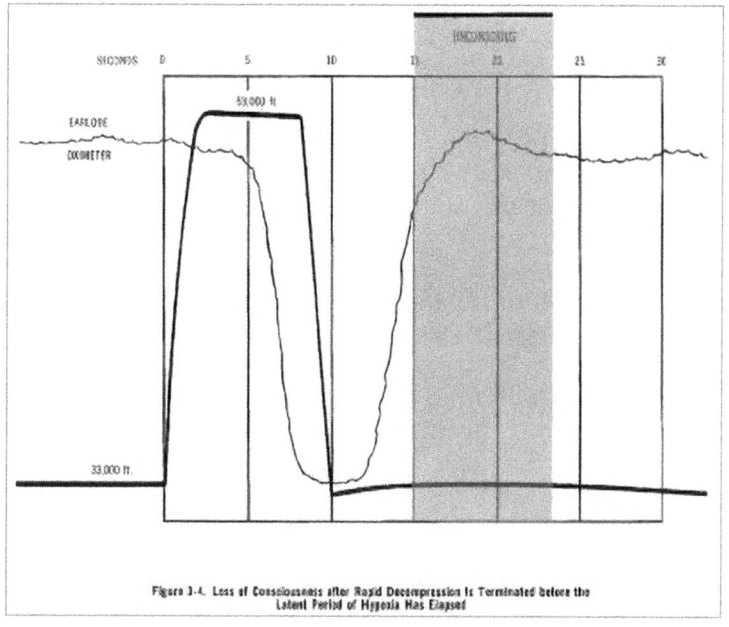

Figure 3-2. Flight Altitude in Thousands of Feet

Figure 3-4. Loss of Consciousness after Rapid Decompression is Terminated before the Latent Period of Hypoxia Has Elapsed

At sea level, arterial blood is normally 93-98% saturated with oxygen, meaning oxygen availability does not significantly regulate respiration. Instead, carbon dioxide tension and arterial blood pH play the predominant chemical role in controlling breathing. The S-shaped oxyhemoglobin dissociation curve illustrates that at the high arterial oxygen tensions typically found at sea level, even a considerable drop in oxygen tension results in only a minor decrease in arterial oxygen saturation. Conversely, arterial carbon dioxide tension is directly proportional to alveolar carbon dioxide tension, making small fluctuations in CO2 sufficient to induce appropriate ventilatory changes before arterial oxygen saturation is significantly affected.

However, above 10,000 feet, the reduced oxygen tension in inspired air makes hypoxia a direct stimulus for respiration. At higher altitudes, a low alveolar oxygen tension means even minor changes in tension can cause a substantial drop in arterial oxygen saturation, independent of carbon dioxide tension changes. This oxygen deficit reflexively stimulates the respiratory center via chemoreceptors, leading to increased respiration. This compensatory hyperventilation, while increasing oxygen intake, also causes excessive carbon dioxide elimination, resulting in hypocapnia. Clinically, this manifests similarly to hyperventilation in anxious individuals, characterized by light-headedness, palpitations, and paresthesias in the extremities and perioral area. Prolonged hyperventilation can lead to unconsciousness and/or carpopedal spasm, with individuals often unaware of overbreathing despite a sensation of stifling. A decrease in arterial carbon dioxide tension causes significant shifts in blood pH, resulting in respiratory alkalosis, which can lead to decreased cerebral blood flow and altered EEG patterns. Hyperventilation during flight is a considerable concern, as the lowering of arterial pCO_2 tends to decrease ventilation, limiting the body's compensatory efforts for hypoxia and introducing undesirable effects.

The pattern of respiration also changes with activity. In a resting individual, inhalation is an active process, while exhalation is largely passive. During exercise, both the rate and volume of respiration change, as does the pattern. For a resting individual, instantaneous inspiratory flow rate increases from zero to 20-30 liters per minute near mid-inspiration, returning to zero at the end. A moderately exercising individual may have a minute respiratory volume of 25-45 liters per minute and instantaneous flow rates as high as 65-90 liters per minute. Maximal inspiratory or peak flow rates can be approximated by multiplying the minute respiratory volume (at STP) by 3.7 for a subject at rest and by 2.8 for an exercising individual.

Understanding Hypoxia: Types and Symptoms

Hypoxia is an acute syndrome in aviation resulting from inadequate tissue oxygenation due to decreased partial pressure of oxygen in the inspired air. While "anoxia," literally meaning "without oxygen," is sometimes used, "hypoxia" (denoting a deficiency rather than a total lack of oxygen) is more accurate, as tissues are rarely entirely devoid of oxygen even in acute altitude sickness.

Hypoxia is generally classified into four distinct types: **1. Hypoxic Hypoxia:** This is caused by a decrease in O_2 pressure in the inspired air or lungs, or by conditions that prevent or interfere with oxygen diffusion across the alveolar membrane. Examples include: * Reduced atmospheric pressure and consequently reduced alveolar pO_2 at altitude. * Interference with respiration due to conditions like asthma (bronchiolar constriction impedes ventilation), pneumonia (fluid in alveoli hinders oxygen diffusion), or obstruction of air passages by tumors or strangulation. * Arterial venous shunts, as seen in congenital cardiovascular conditions. **2. Hypemic (Anemic) Hypoxia:** Characterized by a reduction in the blood's capacity to carry sufficient oxygen due to decreased hemoglobin content. A healthy individual can transport 20 cc of oxygen per 100 cc of blood (1 Gm of hemoglobin carries 1.34 cc O_2). If blood loss reduces hemoglobin by half, oxygen transport capacity also halves to 10 cc per 100 cc. In such cases, tissues may receive insufficient oxygen even if the blood is fully saturated, and cyanosis (requiring over 5 Gm of reduced hemoglobin per 100 cc of blood in skin capillaries) may not be present. Carbon monoxide, nitrites, and sulfa drugs can induce this type of hypoxia by forming stable compounds with hemoglobin, thereby reducing the amount available for oxygen transport. **3. Histotoxic Hypoxia:** Occurs when the body tissues' ability to utilize oxygen is compromised. Substances like alcohol, narcotics, and poisons such as cyanide interfere with cellular oxygen metabolism, even when oxygen supply is normal. During histotoxic hypoxia, venous O_2-hemoglobin saturation is higher than normal because oxygen is not being effectively unloaded and utilized by the tissues. **4. Stagnant Hypoxia:** Results from inadequate blood circulation, despite adequate oxygen-carrying capacity of the blood. Conditions contributing to stagnant hypoxia include heart failure, arterial spasm, blood vessel occlusion, and venous pooling, such as that experienced during positive G maneuvers.

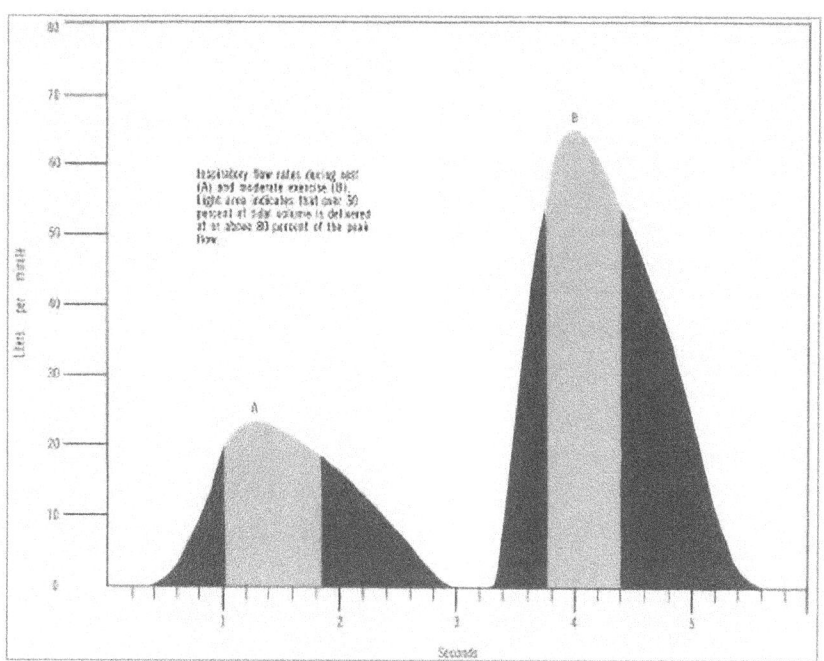

| Stage | Altitude in feet | | Arterial oxygen saturation percent |
	Breathing air	Breathing 100% oxygen	
Indifferent	0-10,000	34,000 to 39,000	95 to 90
Compensatory	10,000-15,000	39,000 to 42,500	90 to 80
Disturbance	15,000-20,000	42,500 to 44,800	80 to 70
Critical	20,000-23,000	44,800 to 45,500	70 to 60

All these forms of hypoxia can pose significant challenges in flight. However, the most prevalent and critical type encountered is hypoxic hypoxia, arising from breathing air with a low partial pressure of oxygen, leading to the "mountain" or "altitude" sickness syndrome.

The appearance and severity of symptoms in acute hypoxic hypoxia depend on several variables: * Absolute altitude * Rate of ascent * Duration at altitude * Ambient temperature * Physical activity * Individual factors (inherent tolerance, physical fitness, emotionality, acclimatization)

While higher altitudes generally lead to more pronounced symptoms, rapid ascent rates can allow individuals to reach higher altitudes before severe symptoms appear. Prolonged exposure significantly impacts symptom onset. High temperatures and physical exertion accelerate symptom development at lower altitudes. Physical fitness and acclimatization from high-altitude residence improve an individual's "ceiling," whereas apprehension and inadequate physiological compensation by respiratory and circulatory systems reduce it.

Hypoxia symptomatology is commonly divided into stages: **1. Indifferent Stage:** The only discernible adverse effect is on dark adaptation, which manifests at altitudes as low as 5,000 feet, underscoring the need for oxygen during night flights, particularly for fighter pilots. Electrocardiographic changes and increased pulse rate may also occur at these low altitudes. **2. Compensatory Stage:** Physiological mechanisms provide some defense against hypoxia. Effects remain latent unless exposure is prolonged or exercise is undertaken. Respiration may deepen or slightly increase in rate. The pulse rate, systolic blood pressure, circulation rate, and cardiac output all increase. **3. Disturbance Stage:** Physiological compensations become insufficient to provide adequate tissue oxygenation, and latent oxygen deficiency becomes overt. Subjective symptoms often include fatigue, lassitude, somnolence, dizziness, headache, breathlessness, and euphoria. Occasionally, unconsciousness can occur without preceding subjective sensations. Objective symptoms include: * **Special Senses:** Impaired peripheral and central vision, diminished visual acuity, weak and incoordinate extraocular muscles, and decreased range of accommodation. Touch and pain sensations diminish or are lost. Hearing is typically one of the last senses to be impaired. * Mental Processes: Intellectual impairment is an early and significant sign, preventing self-awareness of disability. Thinking is slowed, and complex calculations become unreliable. Memory, especially for recent events, is faulty. Judgment is poor, and reaction time is delayed. * **Personality Traits:** Release of basic personality traits and emotions, similar to alcoholic intoxication. Euphoria, elation, pugnaciousness, overconfidence, or moroseness may be observed. * **Psychomotor Functions:** Decreased muscular coordination, making delicate or fine motor movements impossible. This manifests as stammering, illegible handwriting, and poor coordination in tasks like aerobatics or formation flying. * **Hyperventilation Syndrome:** (Discussed in Regulation of Breathing) * Cyanosis: A bluish discoloration of the skin and mucous membranes. **4. Critical Stage:** Characterized by loss of consciousness. This can result from circulatory failure ("fainter"), more common in prolonged hypoxia, or central nervous system failure ("nonfainter," with maintained blood pressure), more common in acute hypoxia. Both types may be accompanied by convulsions and eventual respiratory center failure.

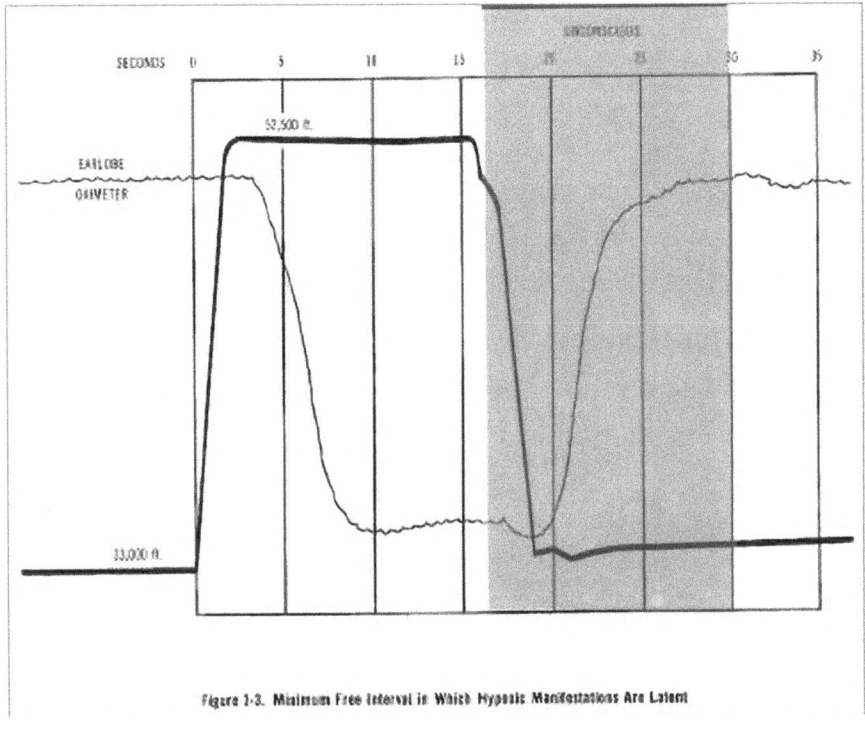

Figure 1-3. Minimum Free Interval in Which Hypoxic Manifestations Are Latent

Post-hypoxic symptoms can include headache and lethargy after prolonged severe hypoxia. Headaches are generally distributed, but particularly acute in the frontal region. Sleep is the best remedy, but 100% oxygen administration is advisable for severe cases. These symptoms are attributed to tissue edema, especially cerebral tissues, resulting from increased capillary permeability caused by hypoxia. Nausea, vomiting, and severe prostration may also occur, typically resolving within 24-48 hours. Permanent cerebral damage from hypoxia is rare, with few authenticated cases.

Individual tolerance to hypoxia varies significantly. This variation in "ceiling" is largely due to the adequacy of physiological adjustments, particularly in breathing. Deeper breathing increases oxygen pressure in the lungs and blood alkalinity, which enhances oxygen uptake by hemoglobin. At extreme altitudes like 40,000 feet, where 100% oxygen is required, alveolar pressure is the sum of partial pressures of water vapor, carbon dioxide, and oxygen. Water vapor pressure is constant (corresponding to saturation at 37°C), so lowering pCO_2 through deep breathing directly increases alveolar pO_2. Inexperienced individuals are more prone to collapse at intermediate altitudes than experienced ones, largely due to psychogenic factors. While hypoxia-induced hyperventilation usually causes only minor symptoms like dizziness, apprehension can lead to more extreme hyperventilation and hypocapnia, exacerbated by splanchnic vasodilation from fear, potentially leading to collapse.

Mitigating Hypoxia: Pressure Breathing and Oxygen Systems

The treatment for hypoxia involves immediate administration of 100% oxygen via inhalation. If respiration has ceased, artificial respiration combined with 100% oxygen is indicated. Persistent peripheral circulatory failure requires specific diagnosis and treatment. Preventing hyperventilation in flying personnel is primarily achieved through indoctrination on the proper use of USAF oxygen equipment. Recovery from hypoxia is rapid with sufficient oxygen; an individual near unconsciousness can regain full faculties within 15 seconds. Deep breathing of oxygen may cause a momentary flash of dizziness, followed by complete restoration of normal function.

To prevent hypoxia above 40,000 feet, methods for increasing alveolar oxygen partial pressure are necessary, with positive pressure breathing being a key technique. This involves an oxygen system delivering 100% oxygen at pressures exceeding ambient. Specialized pressure breathing regulators, masks, and mask valves are essential.

In **continuous positive pressure breathing**, the mechanics of respiration become more challenging at higher pressures. The normal pattern of active inspiration and passive expiration reverses, becoming passive inspiration and very active expiration. **Intermittent positive pressure breathing** offers partial compensation by delivering oxygen under pressure only during inspiration, making both inspiration and expiration passive. However, this often leads to symptomatic hyperventilation, a significant disadvantage.

The **mean mask pressure** of oxygen, the average pressure at the mask over a complete respiratory cycle, is crucial for physiological protection against hypoxic hypoxia. Continuous positive pressure breathing provides a mean mask pressure nearly equivalent to the regulator's delivery pressure, offering superior protection compared to intermittent positive pressure breathing, which delivers only one-third to one-half of the peak mask pressure. The effectiveness of continuous positive pressure is also influenced by subjective response. A rapid inspiration followed by prolonged expiration increases mean mask pressure but also raises intrathoracic pressure, restricting blood flow through the lungs and increasing venous pressure, potentially reducing cardiac output and causing blood pooling in the lower extremities and abdomen. Conversely, a prolonged inspiration followed by rapid expiration lowers mean mask pressure slightly but significantly reduces average intrathoracic pressure, allowing higher oxygen pressures without adverse circulatory effects.

Pressure breathing at 15-30 mm Hg (8-15 inches of water) can be sustained for limited periods. Above 30 mm Hg, subjects may experience fatigue, overinflation, and congestion in the frontal sinuses. Higher pressures can cause pain in the ears and posterior pharynx due to overdistension, often leading to termination of attempts at 60 mm Hg. At 60-100 mm Hg, parenchymal lung damage from overexpansion is likely without counter-pressurization. The most significant limitation of pressure breathing is its impact on the cardiovascular system. Increased intrathoracic pressure compresses lung tissue, resisting blood flow and increasing venous pressure, which leads to decreased cardiac output and blood pooling. This blood displacement can cause loss of consciousness.

Therefore, for flights exceeding 40,000 feet, individual pressurization via a counter-pressurization suit or a pressurized cabin is required. A pressurized cabin allows flight up to 50,000 feet without continuous oxygen equipment, provided cabin pressure is maintained below 10,000 feet (523 mm Hg). However, Air Force regulations prohibit flying above 50,000 feet without a counter-pressurization suit, regardless of cabin altitude. Cabin pressurization involves a compromise between mechanical and physiological constraints, meaning the pressure differential cannot always create ideal conditions. Thus, at higher aircraft altitudes, cabin altitude also increases. Maintaining cabin altitude below 40,000 feet avoids the need for positive pressure breathing and reduces the incidence of decompression sickness. Ideally, cabin altitude below 34,000 feet is sought, as breathing 100% oxygen at this level is equivalent to breathing air at sea level.

A major risk of pressurized cabins is rapid decompression due to mechanical failure or combat damage, exposing aircrews to critical altitudes. Between 40,000 and 50,000 feet, pressure breathing regulators provide sufficient pressure to maintain consciousness during an emergency descent. Above 50,000 feet, pressure breathing alone is inadequate due to the high positive pressures required. Consequently, all pilots and crew flying above 50,000 feet must wear counter-pressurization garments that automatically inflate upon loss of cabin pressurization, ideally allowing mission completion. Sudden exposure to critical altitudes (e.g., above 50,000 feet) leaves only 20-30 seconds between decompression and loss of consciousness, underscoring the necessity of familiarization with oxygen equipment and pressure suits.

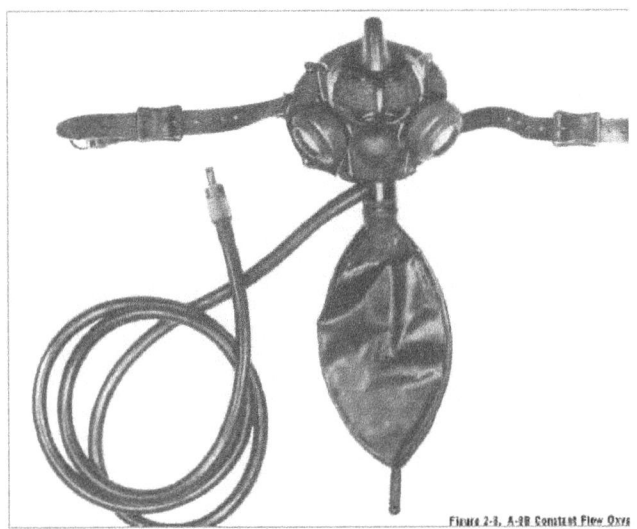

Figure 2-3. A-2B Constant Flow Oxygen

Oxygen Equipment for Flight

The evolution of Air Force oxygen equipment has progressively extended man's altitude ceiling, highlighting the essential collaboration between engineers and physiologists. Generally, an aircraft oxygen system comprises oxygen storage containers, tubing for delivery, a metering device to control flow, and a mask to direct oxygen to the respiratory system.

Early USAF equipment for hypoxia protection was a simple tube connected to a high-pressure oxygen cylinder (up to 1,800 psi) via a valve-type metering device. A pipestem delivered a continuous flow of oxygen to the user's mouth. This system improved conditions up to about 20,000 feet but had several drawbacks: it didn't protect normal nose breathers, oxygen was wasted during expiration (about half the total time), and the pipestem was uncomfortable to hold, especially in unheated cockpits.

Continuous Flow Oxygen Systems The need for economy in weight and space in military aircraft, along with increasing altitude ceilings, drove the development of systems that could regulate oxygen flow more precisely. This led to the continuous flow oxygen system, employing a lightweight oronasal mask with a re-breather bag (approx. 800 cc capacity). The re-breather bag captures the initial portion of exhaled gas (from respiratory dead space), which is primarily unused oxygen, allowing for its reuse. It also acts as a reservoir for oxygen flowing from cylinders during the expiratory phase. The mask includes sponge-rubber discs functioning as exhalation valves and inspiratory ports for ambient air intake at lower altitudes when the re-breather bag and oxygen flow are insufficient.

Oxygen flow is controlled by a metering device or regulator (e.g., A-1I, which is automatic, or manual versions). These regulators deliver oxygen at constant pressure regardless of cylinder pressure changes. The pilot adjusts a calibrated valve to match flight altitude, ensuring appropriate supplementary oxygen. While a continuous flow system can theoretically supply 100% oxygen, its effective upper limit is 40,000 feet under ambient pressure, providing conditions similar to breathing air at 10,000 feet. This applies primarily to conditions of rest or mild activity. For passengers in transport aircraft, it is adequate. However, for active military aircrews, higher oxygen demands or mask leakage (especially at higher altitudes) can lead to hypoxia. Due to practical experience, an arbitrary ceiling of 25,000 feet (7.62 Km) was established for this equipment type. Continuous flow systems are simple, reliable, and common in cargo and transport aircraft, suitable for moderate altitudes. The inherent risk of even small amounts of ambient air entering the respiratory passages above 25,000 feet necessitated solutions for inboard mask leakage.

Demand-Type Oxygen Systems An advancement over continuous flow, the demand-type oxygen system delivers oxygen only during the inspiratory phase of the breathing cycle. Extensive tests confirmed that this system ensures 100% oxygen delivery at high altitudes, provided an airtight face-to-mask seal is maintained. Wartime pressures expedited the development and standardization of this system for combat aircraft.

The standard demand oxygen mask is a simple mechanism designed for a comfortable facial fit. It features a single flapper valve in the mask facepiece base, which allows expired air to escape to the atmosphere. During inspiration, the flapper seals tightly, preventing ambient air entry. The oxygen flow is regulated by a diluter-demand regulator, a relatively simple mechanism responding to normal pressure changes during breathing. It consists of a round box with a rubber diaphragm attached to a valve arm. Inhaling creates a slight negative pressure within the regulator, drawing the diaphragm inward, opening a port to the pressure reduction stage, and allowing oxygen to flow. The initial chamber reduces cylinder pressure to a constant, fixed pressure behind the diaphragm-operated valve.

The primary point of air leakage in this system is the face-to-mask seal. Despite various mask designs aimed at effective sealing, minute leakage is often inevitable. For safety, this limits the demand oxygen system to altitudes below 35,000 feet (10.67 Km). Oxygen conservation is achieved by mixing ambient air with oxygen up to 34,000 feet (10.36 Km). A diluter mechanism, using metal bellows connected to air and oxygen valves, adjusts the mixture. As altitude increases, trapped air in the bellows expands, gradually increasing oxygen valve opening and closing the air inlet, leading to 100% oxygen delivery at 34,000 feet. A dilution lever allows the user to manually switch to "100 per cent oxygen" at any altitude. For mild hypoxia, an emergency valve on the demand regulator provides a rapid, continuous flow of oxygen, capable of reviving a hypoxic individual within seconds and maintaining normal arterial oxygen saturation even during moderate exercise-equivalent ventilation.

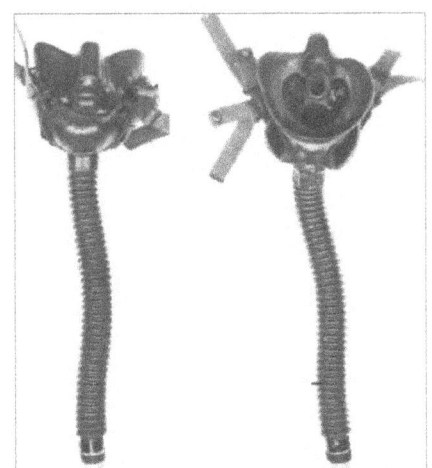

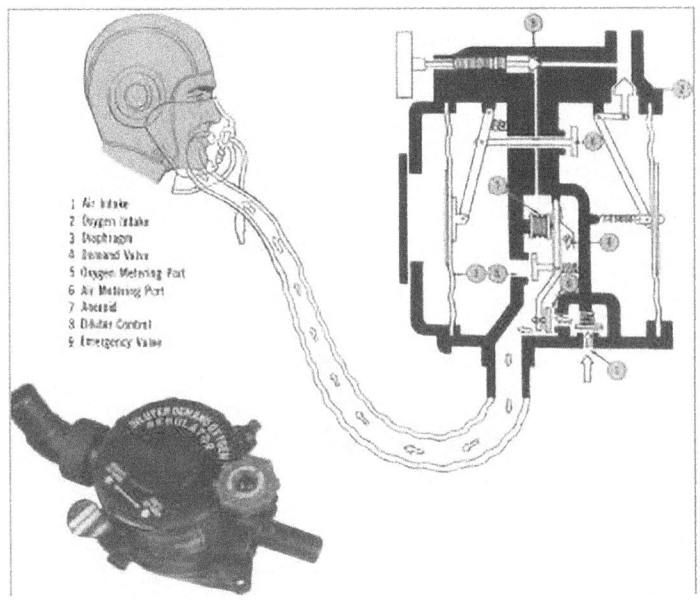

1 Air Intake
2 Oxygen Intake
3 Diaphragm
4 Demand Valve
5 Oxygen Metering Port
6 Air Metering Port
7 Aneroid
8 Diluter Control
9 Emergency Valve

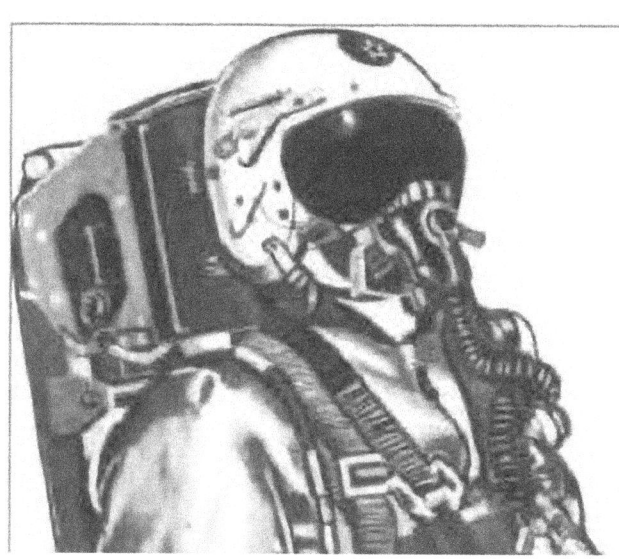

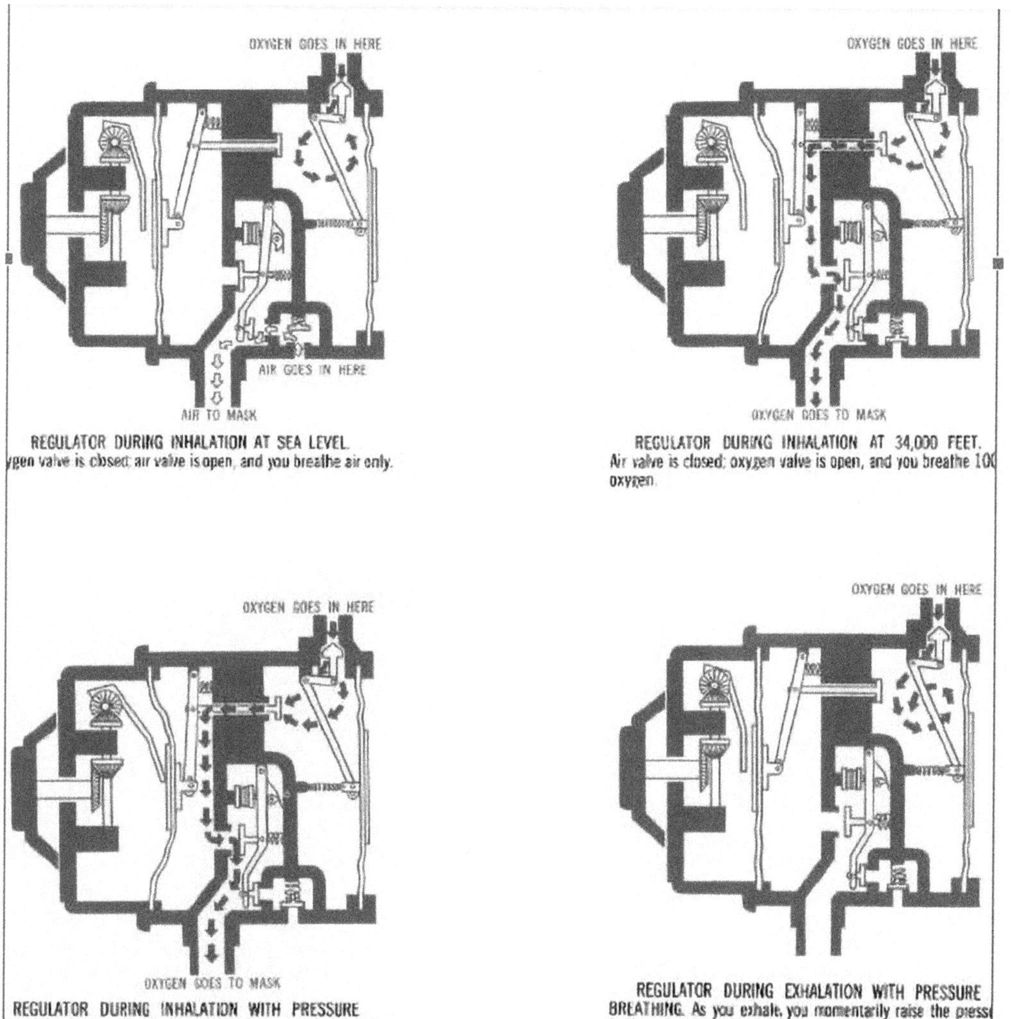

OXYGEN GOES IN HERE

REGULATOR DURING INHALATION AT SEA LEVEL.
ygen valve is closed: air valve is open, and you breathe air only.

AIR GOES IN HERE

AIR TO MASK

OXYGEN GOES IN HERE

REGULATOR DURING INHALATION AT 34,000 FEET.
Air valve is closed: oxygen valve is open, and you breathe 100
oxygen.

OXYGEN GOES TO MASK

OXYGEN GOES IN HERE

REGULATOR DURING INHALATION WITH PRESSURE
EATHING. Spring presses down on diaphragm, opening de
d valve, and forcing oxygen into the mask under pressure

OXYGEN GOES TO MASK

OXYGEN GOES IN HERE

REGULATOR DURING EXHALATION WITH PRESSURE
BREATHING. As you exhale, you momentarily raise the pressu
forcing the diaphragm up against the spring tension. The
mand valve closes and no oxygen flows.

Pressure-Breathing Systems Maintaining flyers in normal condition above 34,000 feet requires two key provisions: eliminating inboard mask leakage for 100% oxygen delivery up to 40,000 feet, and delivering oxygen at pressures exceeding ambient above 40,000 feet. Positive pressure breathing, by adding sufficient pressure to the mask, sustains a normal alveolar partial pressure of oxygen. Mechanically, modifications to eliminate inboard mask leakage and provide positive pressure are similar. Supplying oxygen at a small positive pressure (about 2 inches of water) prevents ambient air from being drawn in during inspiration, even if the mask seal is broken. This "safety" pressure concept is applied in standard oxygen systems for altitudes between 30,000 and 40,000 feet.

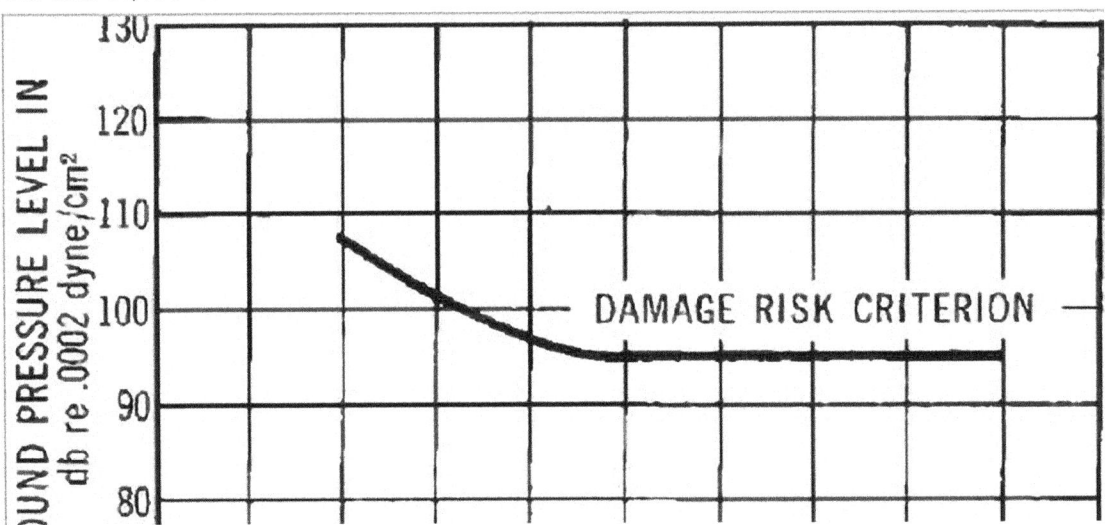

Enhanced mask-to-face seals allow pressure breathing up to 30 mm Hg. The conversion of a demand mask to a pressure-breathing system involves two main changes: 1. The mask is molded with an inner flap that seals against the face when oxygen is delivered at positive pressures. 2. The standard exhalation valve, which opens at low positive pressure, is altered. A direct connection from the incoming oxygen line to the underside of the exhalation valve ensures the oxygen pressure closes the valve. To exhale, a greater force than the incoming oxygen pressure is required. Check valves over the oxygen inlet ports facilitate the necessary exhalation pressure within the mask.

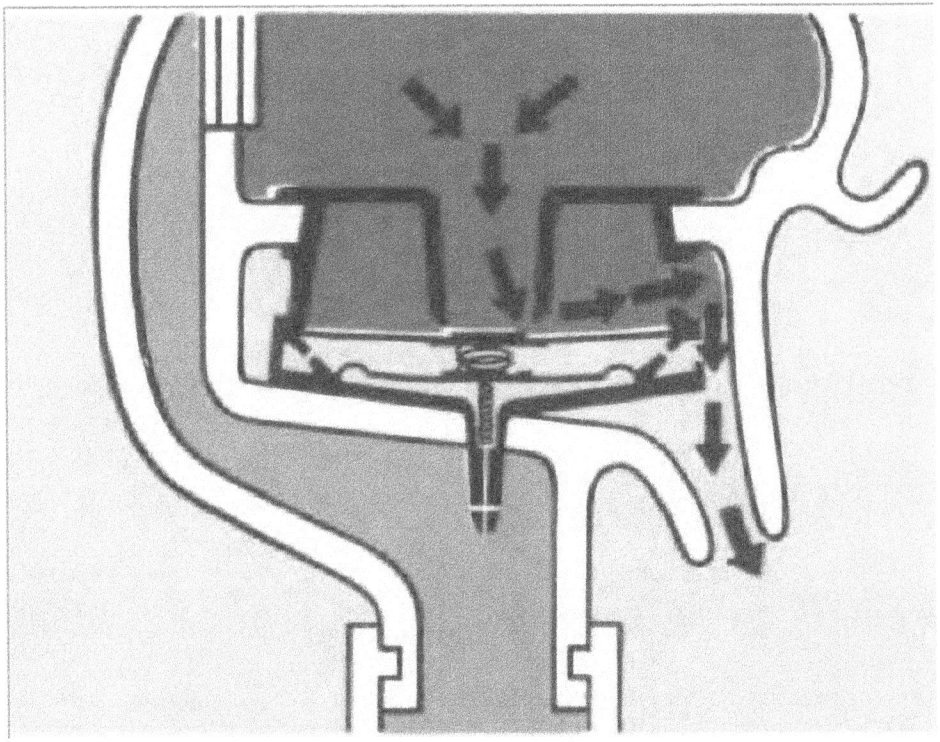

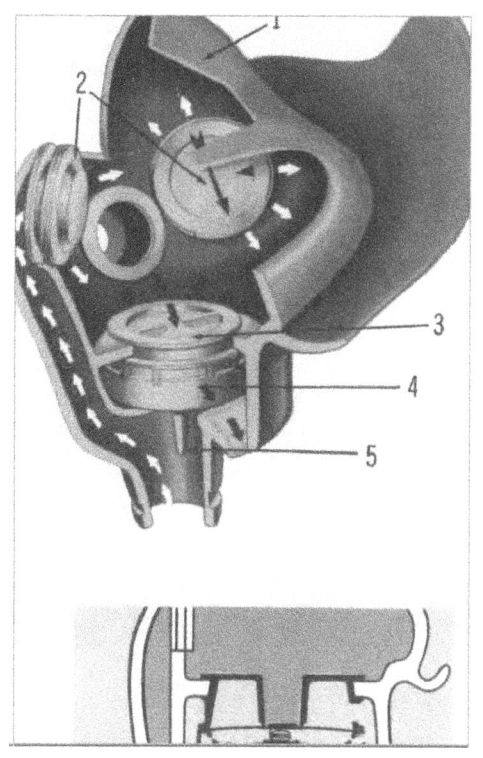

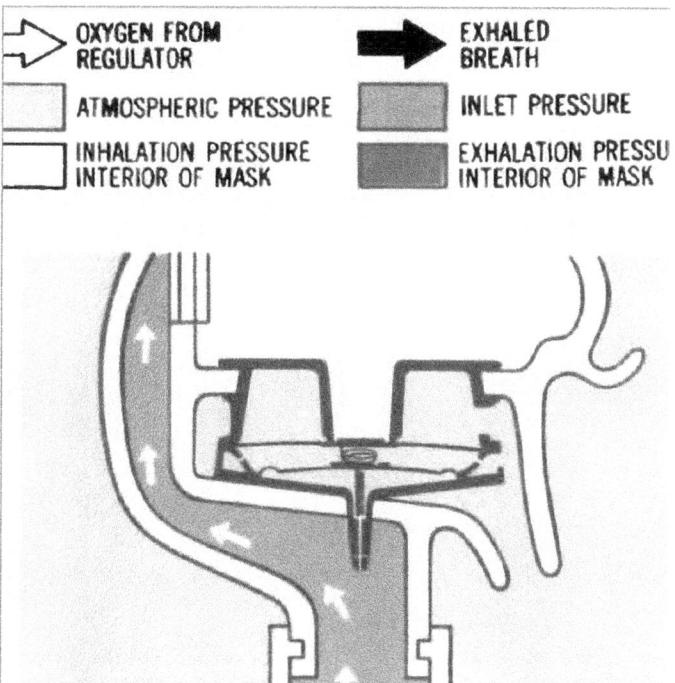

OXYGEN FROM
REGULATOR

EXHALED
BREATH

ATMOSPHERIC PRESSURE

INLET PRESSURE

INHALATION PRESSURE
INTERIOR OF MASK

EXHALATION PRESSU
INTERIOR OF MASK

IF THESE INLET CHECK VALVES ARE NOT SET INTO
PLACE PROPERLY, THE MASK WILL NOT WORK

A spring applied to the diaphragm converts a standard demand regulator into a pressure-demand system. The pressure delivered to the mask is varied manually via a lever on the regulator. For emergency situations in aircraft with operating ceilings up to 50,000 feet, an automatic aneroid mechanism activates the pressure-demand system.

Liquid Oxygen (LOX) Systems Advances in inflight refueling and aircraft pressurization systems have led to increased flight durations and, consequently, a greater demand for crew oxygen supplies. This challenge was addressed by the development of liquid oxygen (LOX) systems. Despite some disadvantages, LOX systems offer substantial advantages by providing tremendous quantities of breathing oxygen in a compact, lightweight unit. A LOX system generally consists of a container, pressure build-up and vent valves, an evaporator coil, check valves, pressure and capacitance type contents gauges, and associated relief valves. For instance, a USAF Type A-3 5-liter liquid oxygen converter has a capacity of 5 liters, operates at 70 psi, and weighs 14.0 lbs empty (26.5 lbs full).

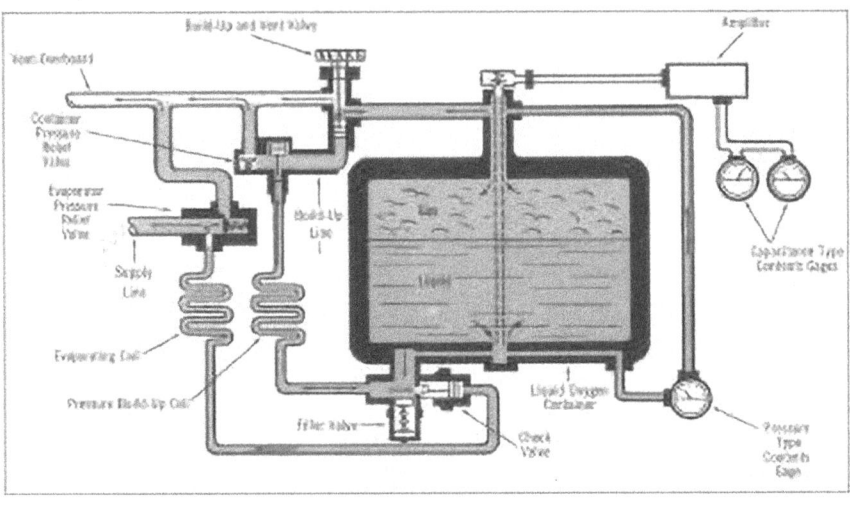

Figure 2-16. USAF Type A-3 5-Liter Liquid Oxygen Converter

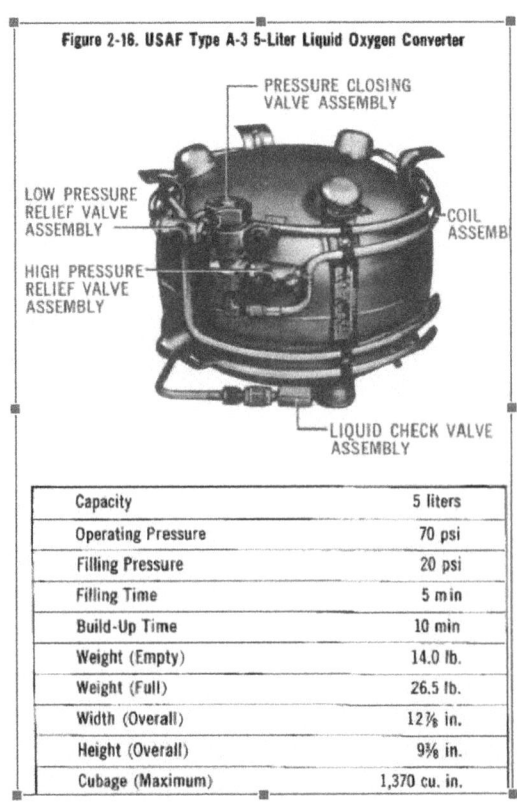

Capacity	5 liters
Operating Pressure	70 psi
Filling Pressure	20 psi
Filling Time	5 min
Build-Up Time	10 min
Weight (Empty)	14.0 lb.
Weight (Full)	26.5 lb.
Width (Overall)	12⅞ in.
Height (Overall)	9⅜ in.
Cubage (Maximum)	1,370 cu. in.

References

USAF. (2021). "Physiology of Flight (AFP 160-10-4)". U.S. Air Force.

USAF School of Aviation Medicine. (2018). "Handbook of Respiratory Physiology" (Project 21-2301-0003).

USAF. (2024). "Your Body in Flight (AFP 160-10-3)". U.S. Air Force.

Armstrong, H. G. (2016). "Aerospace Medicine" (Chapters 8, 9, 10, 13). Williams and Wilkins Co.

Benson, O. O. Jr., & Strughold, H. (2020). "Physics and Medicine of the Atmosphere and Space". John Wiley & Sons Inc.

Chapter 5: Understanding Human Response to Accelerative Forces in Flight

Flight, inherently an unnatural human endeavor, exerts its most profound impacts on the body through the accelerative forces encountered during aerial maneuvering. While there are no human limitations to speed in straight and level flight, constraints arise from changes in velocity or direction. A comprehensive understanding of accelerative forces and their relationship to the human body in flight is foundational to the practice of aviation medicine. This chapter will delve into the effects of accelerative forces, encompassing equilibrium, spatial orientation, airsickness, and G-tolerance.

Fundamentals of Aircraft Motion and Acceleration

All flying is predicated upon fundamental maneuvers, most of which involve movements around distinct axes of the aircraft. An aircraft rotates about three primary axes: the lateral, vertical, and longitudinal. These rotations are controlled by three corresponding flight controls: the elevators, rudder, and ailerons.

The lateral axis is an imaginary line extending from wing tip to wing tip, passing through the aircraft's center of gravity and perpendicular to the longitudinal and vertical axes. Rotation around this axis, known as pitch, is controlled by the elevators. These are movable horizontal surfaces on the tail of the aircraft, manipulated by forward or backward pressure on the control stick. For instance, in straight-and-level flight, forward pressure lowers the nose, while backward pressure raises it.

The vertical axis is another imaginary line that runs through the center of gravity, perpendicular to both the lateral and longitudinal axes. Rotation about this axis, termed yaw, is governed by the rudder. Pressure applied to the right rudder moves the nose to the right, and vice-versa for the left rudder.

The longitudinal axis is an imaginary line traversing the aircraft from nose to tail, passing through the center of gravity and perpendicular to the lateral and vertical axes. Rotation around this axis, referred to as roll, is controlled by the ailerons. Ailerons are movable panels located on the outer trailing edge of the wings, operated by side pressure on the control stick. Rotation is induced by a lift differential: a raised aileron on one wing decreases lift, causing that wing to descend, while a lowered aileron on the opposite wing increases lift, causing it to ascend. Moving the control stick toward a wing raises its aileron surface, initiating a roll in that direction.

The magnitude of pressure exerted on a control surface is dictated by airspeed and the extent to which the surface is moved from its streamlined position. At higher airspeeds, even minor control movements result in more abrupt changes in aircraft attitude. Beyond these primary rotations, aircraft may also experience other motions like "bumping" (rapid vertical movements in turbulent air) or "corkscrewing" (oscillating movements of the tail in larger aircraft), or any combination of yawing, pitching, and rolling. Any alteration in aircraft attitude invariably involves acceleration, or a change in velocity.

The fundamental concepts of motion trace back to Sir Isaac Newton's three laws of motion, expressed in 1687, which elucidate the nature of motion and its causative forces:

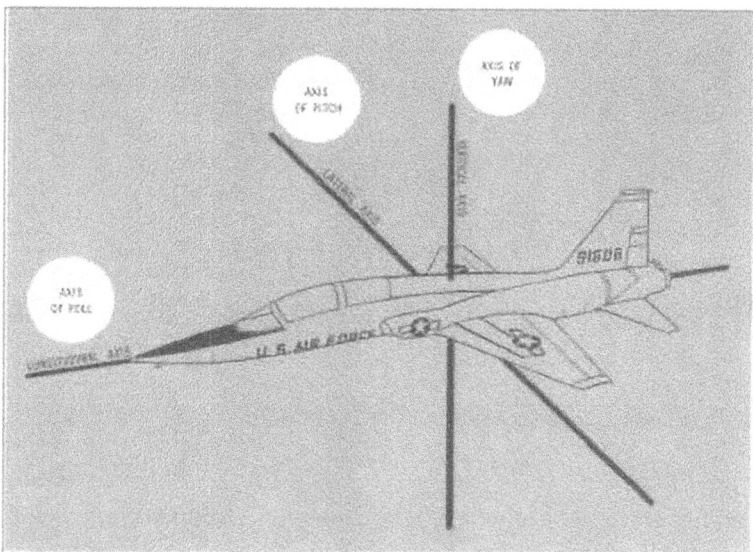

Newton's first law, the Law of Inertia, states that a body at rest tends to remain at rest, and a body in motion tends to remain moving at the same speed and in the same direction. This implies that an external force is required to initiate or halt motion. Once an aircraft is airborne, inertia works to maintain its motion, subject to external forces. Similarly, a pilot's body possesses inertia, tending to continue its motion path during maneuvers, such as being pressed against the seat during a pull-up from a dive.

Newton's second law, the Law of Acceleration, addresses the force involved in overcoming inertia. Acceleration is defined as the change in velocity per unit of time, encompassing changes in speed (acceleration and deceleration) and direction. The law states that when a body is acted upon by a constant force, its resulting acceleration is inversely proportional to the mass of the body and directly proportional to the applied force. Mathematically, this is expressed as $a=F/M$ or $F=Ma$, where F is force in pounds, M is mass, and a is acceleration in feet per second squared. Mass is a constant quantity, distinct from weight, which varies with gravity.

Newton's third law, the Law of Action and Reaction, posits that for every action, there is an equal and opposite reaction. A prime example in modern aviation is jet propulsion, where the rapid expulsion of hot gases from the turbine's tailpipe generates an equal and opposite reaction that propels the aircraft forward.

" G " Forces The most universally recognized acceleration is that due to gravity, which is 32.2 feet/second^2. The force producing this acceleration is termed 1G. Consequently, an acceleration of 640 feet/second^2 is equivalent to 20 G's, being twenty times the acceleration of gravity. Force and acceleration are directly proportional ($F = Ma$).

Types of Acceleration Acceleration, defined as the rate of change in velocity in G units, manifests in different forms during flight. The relationship $a=V^2/r$ (where V is airspeed and r is the radius of turn) is fundamental to understanding how G forces develop in flight; doubling airspeed, for instance, quadruples G's.

Linear acceleration results from a change in speed while moving in a straight line, as when an aircraft increases speed from 200 to 300 mph. It is also experienced in crash landings, catapult takeoffs, parachute openings, and landing shocks. Linear G can be calculated by the formula: Linear $G = [V_2^2 - V_1^2] / (32 \times 2d)$, where V_1 is initial speed, V_2 is final speed, and d is the distance over which acceleration occurs. Positive values indicate acceleration, and negative values deceleration.

Radial acceleration occurs with any change in direction at a constant speed, such as going around a curve or executing a loop in an aircraft. Radial G is calculated as: Radial $G = V^2 / (32 \times r)$, where V is speed in feet/second and r is the radius of the turn in feet.

Angular acceleration involves simultaneous changes in both speed and direction, exemplified by an aircraft in a tight spin. The forces in such a maneuver can be so intense that a pilot may be unable to escape. Angular acceleration can be calculated by combining the linear and radial formulae. It is important to note that the physical and physiological effects of linear, radial, or angular acceleration are identical when the G qualities are the same.

The effects of G forces are influenced by several factors: (1) the **degree (intensity)** of the force; (2) the **time (duration)** of application; (3) the **rate** of application; (4) the **area and site** on the body over which the force is applied; and (5) the direction of the accelerative force relative to the body's long axis. Generally, greater intensity leads to more severe effects, but duration is also crucial; high G forces for extremely short periods can be tolerated as well as low G forces for longer periods. A higher rate of application typically results in more severe effects, as seen in crash landings where stopping distance impacts the rate of acceleration. Distributing a given force over a larger body area reduces its harmfulness, and the specific site of application (e.g., head vs. other body parts) significantly alters the outcome. Finally, the direction of a prolonged accelerative force profoundly determines the physiological responses, as the centrifugal force increases the effective weight of the body and its components.

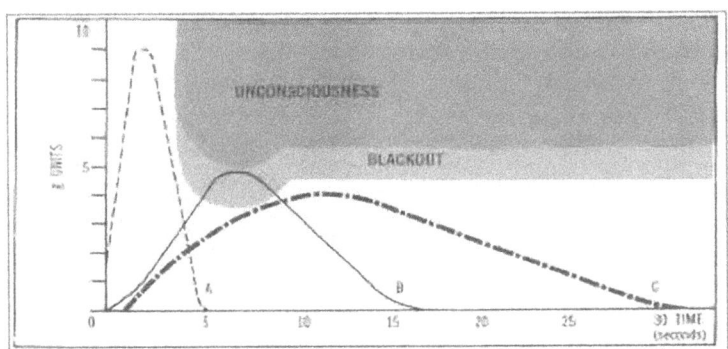

The direction of G force application to the body is of paramount physiological interest due to varied effects. Positive G occurs when the force acts in a head-to-foot direction, akin to standing upright. It causes a temporary caudal displacement of blood, potentially leading to blackout and unconsciousness. Negative G forces act from foot to head, as when inverted. These induce a temporary displacement of blood towards the head and neck, causing "red out" and potentially unconsciousness. Transverse G refers to forces acting on the body in a prone or supine position. Humans are most tolerant to transverse G, enduring higher magnitudes and longer durations compared to positive or negative G. This type of force causes transverse blood displacement, leading to labored respiration and eventual unconsciousness from prolonged exposure. Differentiating these G-force directions is crucial because their physiological consequences are distinct; for example, an average pilot can withstand 4-6 positive G's for 3-5 seconds without blacking out, but only 3 negative G's before risking "red out" and unconsciousness, yet can endure up to 15 transverse G's with only moderate discomfort.

Physiological Impacts of G-Forces (Positive, Negative, Transverse)

The physiological impacts of G-forces are diverse and depend heavily on the direction and magnitude of the force. These forces can significantly compromise bodily functions, with particularly notable effects on the cardiovascular system and sensory perception.

Effects of Positive G Positive G-forces manifest their effects predominantly in three areas: the body as a whole, the viscera, and the cardiovascular system, with the latter being the most critical. First, regarding the body, during a positive G maneuver, the body's effective weight increases in direct proportion to the G magnitude. A 200-pound person, for instance, would effectively weigh 800 pounds under a 4 G maneuver. This severely curtails normal activities; the flyer is pressed into the seat, limbs feel heavy, cheeks sag, and free movement becomes nearly impossible. A maneuver exceeding 2 to 3 G's (regardless of direction) can immobilize a pilot, highlighting the necessity of ejection seats for escape from spinning aircraft. Second, the viscera are pushed caudally during positive G maneuvers. The increased weight of the viscera pulls the diaphragm downward, augmenting the relaxed thoracic volume and disrupting the mechanics of respiration. Third, the cardiovascular system is particularly vulnerable. Due to human anatomy, with the heart positioned roughly at the junction of the upper and middle thirds of the cylindrical body, and the head (the most blood pressure-sensitive structure) about 30 cm above the heart when seated, positive G-forces can critically impair cerebral blood flow. A force of 5 positive G's, for example, creates a 120 mm Hg hydrostatic pressure from a 30 cm blood column. This pressure can counteract normal arterial systolic pressure (around 120 mm Hg), causing cerebral blood perfusion to cease and resulting in unconsciousness. At approximately 4 G's, visual dimming or "grayout" occurs as static intra-ocular pressure (around 20 mm Hg) causes retinal arteries to collapse when systolic arterial pressure in the head drops to this level, impairing retinal function and narrowing vision from the periphery. Blackout, where vision is completely lost but consciousness is maintained, follows at 4.0 to 4.5 G's. At 4.5 to

5.0 G's, cerebral blood flow ceases, leading to unconsciousness. The typical sequence is dimming of vision, then blackout, and finally unconsciousness. While it was previously thought that blood pooling in the lower body and decreased venous return were primary causes, recent research indicates that for effects occurring in less than 10 seconds, the increased weight of the blood is the main culprit. Decreased venous return, while important, likely contributes more to blackout from lower G-forces (3-3.5 G's) over longer durations (15 seconds or more).

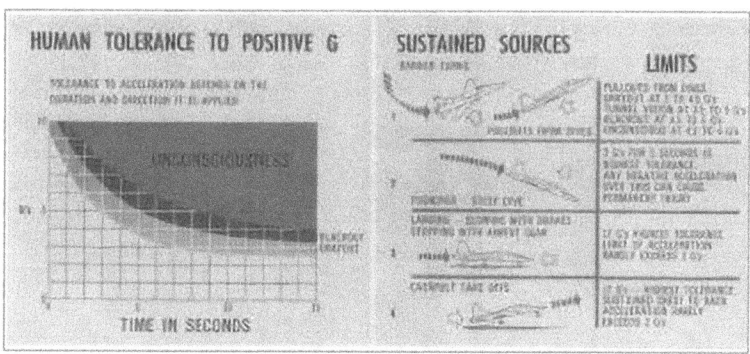

Effects of Negative G Negative acceleration, where force is applied from foot to head, leads to an increased arterial pressure at head level. The veins outside the cranial cavity experience precipitously high pressures, which can rupture thin-walled venules. While intracranial venous pressure rises, it is counteracted by a concomitant rise in cerebrospinal fluid pressure, minimizing actual danger of intracranial hemorrhage. The primary damage from negative G's is typically hemorrhages within the eye. Negative G-forces cause distension of the jugular veins, as well as veins of the sinuses and conjunctiva. There is a sharp rise in both arterial and venous pressure at the head level. A sudden force of 3 negative G's is considered the limit of human tolerance, as it can induce venous pressures of 100 mm Hg, causing small conjunctival bleeding and marked head discomfort. However, the skull-enclosed cerebral vessels, bathed in cerebrospinal fluid, are protected and do not show significant deviation from their normal caliber, meaning cerebral vascular damage is unlikely as long as the skull remains intact. Pilots may experience "red out" during negative G maneuvers. Although rare in experimental settings, this phenomenon in aircraft might be due to the lower eyelid gravitating over the cornea (the muscles of the lower eyelid being relatively weak). The engorgement of eyelid vessels could create the perception of a red curtain obscuring vision. Prolonged negative G can eventually lead to circulatory distress. Blood pooling occurs in the head and neck due to the increased effective weight of the blood, resulting in transudation of fluid from blood into tissue spaces. Inadequate blood return to the heart follows due to effective blood volume loss, leading to blood stagnation in the head and neck. Consequently, the cerebral arteriovenous pressure differential becomes insufficient to maintain consciousness. However, negative G's typically pose less of a problem in military flying because they are uncomfortable for pilots, who tend to avoid such maneuvers.

Effects of Transverse G Humans exhibit greater tolerance to transverse G-forces than to either positive or negative G-forces because transverse G interferes minimally with blood flow. Nevertheless, extreme transverse G values (12 to 15 G's) sustained for relatively long periods can cause organ displacement or a shift in the heart's position, thereby impeding respiration. Human subjects exposed to 15 G's for 5 seconds have reported chest pain and ventricular arrhythmias. Surface petechial hemorrhages have also been observed, likely due to forceful blood pooling in the dependent half of the body.

Enhancing G-Tolerance and Protective Measures

Tolerance to G-loads can vary, but generally, it is possible to enhance it through specific physiological adjustments and protective equipment. The body's capacity to withstand G-forces can be significantly improved by shortening the heart-to-head distance or by increasing systolic blood pressure.

" G " TOLERANCE G-tolerance exhibits individual variation but is relatively constant. Strategies to improve it leverage the aforementioned principles. Leaning forward reduces the length of the blood column to the head, facilitating blood flow from the heart to the neck and head. Tensing all skeletal muscles activates pressure reflexes, thereby increasing blood pressure. This latter technique, known as the M-1 maneuver, also counteracts venous pooling in the lower extremities. It is crucial to maintain an open glottis during the M-1 maneuver, avoiding the Valsalva maneuver (closing the glottis and exhaling against it) as it decreases G-tolerance by reducing venous return and cardiac output. Emotional states, such as rage or fear, can temporarily increase blood pressure and pulse rate, thus boosting G-tolerance. However, this effect is highly variable and may be outweighed by other factors. Physical fitness is a significant determinant of G-tolerance; conversely, poor muscle tone, fatigue, sleep deprivation, hypoxia, hypoglycemia, illness, and excessive alcohol or tobacco consumption all diminish it. Experience also plays a vital role. Seasoned pilots develop compensatory "reflexes," such as muscle tensing during G-onset, and learn to anticipate and recognize their G-limits. They become proficient in employing effective countermeasures, whether leaning forward or tensing muscles.

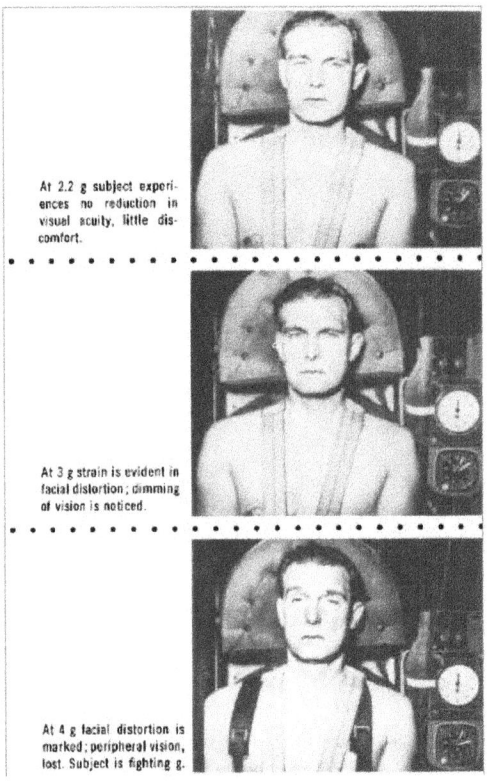

At 2.2 g subject experiences no reduction in visual acuity, little discomfort.

At 3 g strain is evident in facial distortion; dimming of vision is noticed.

At 4 g facial distortion is marked; peripheral vision, lost. Subject is fighting g.

Devices to Protect Against G Forces The primary methods currently used to combat positive G-forces are the M-1 maneuver and the anti-G suit.

M-1 Maneuver Straining maneuvers are universally adopted by experienced fighter pilots, with individual variations in technique. The M-1 maneuver is highly effective, increasing G-tolerance by approximately 1 to 1.5 G's. It involves bending the trunk forward at the hips to achieve postural protection by lowering the head's level relative to the heart, which aids blood flow to the neck and head. Simultaneously, abdominal and chest muscles are contracted, and breath is slowly expelled, with respiratory cycles repeated every 5 to 10 seconds. Arm and leg muscles are also tensed. This maneuver is fatiguing, and its maintenance becomes progressively difficult with increased acceleration duration.

Anti-G Suits External counter-pressure below the heart, achieved through anti-G suits, has proven invaluable in combat aircraft for augmenting human tolerance to G-forces. Initially conceived to balance hydrostatic forces resulting from gravitation, this concept evolved to include garments that compress arteries, thereby increasing arterial pressure. Modern single-pressure and air bladder suits effectively address both requirements by preventing venous pooling and enhancing arterial pressure at heart and head levels through mechanical constriction. This is largely accomplished by a large abdominal bladder that, when inflated during a G maneuver, compresses the viscera, driving blood into the thorax. Additionally, it elevates the diaphragm, substantially reducing the heart-to-head distance. The current standard anti-G suit, USAF type G-4A, features a single pneumatic bladder system integrated into a flying suit. This device boosts tolerance to accelerative forces by about 2 G's, offering protection comparable to the M-1 maneuver. It obviates the need for straining, yet the M-1 can still be employed for further protection if desired. The G-3A suit is a foundational design, providing pressure to the abdomen and major leg muscles via a one-piece bladder system worn over flying clothing. The G-5A capstan suit, conversely, promises greater efficiency and comfort by pressurizing the body surface through tension from tapes encircling distensible high-pressure tubes along the legs. In jet aircraft, anti-G suit inflation is managed by a line connected to the power plant. Air from the compressor is metered to the suit via a special valve that initiates inflation only when acceleration surpasses 2 G's. The pressure then increases proportionally with acceleration. This allows the pilot to relax while the suit compresses the legs and abdomen, effectively substituting for the muscular effort of the M-1 maneuver. The pressure is generally comfortable and can be sustained indefinitely. It is critical to understand that an anti-G suit does not elevate human tolerance beyond the structural stress limits of the aircraft; instead, it optimizes the pilot's capacity relative to the aircraft's capabilities. In aircraft stressed to 5 G's or less, anti-G suits are not strictly necessary unless prolonged maneuvers (0.5 to 1 minute) are anticipated. However, in standard fighter aircraft, anti-G suits are preferred due to the fatiguing, distracting, and unreliable nature of voluntary straining.

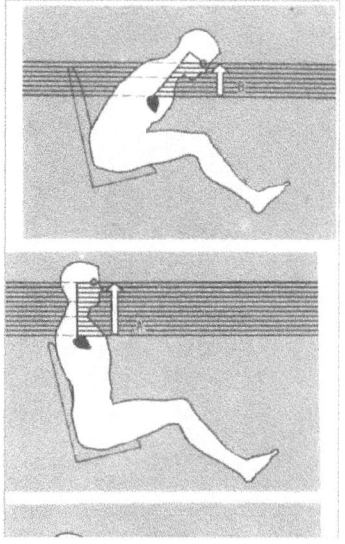

Sensory Perception and Vestibular System in Flight

The ability of an individual to perceive the attitude of their aircraft relative to the Earth's surface is termed aerial equilibration or spatial orientation. Maintaining equilibrium in both static and dynamic states necessitates continuous muscular activity, orchestrated by the central nervous system, which relies heavily on sensory input. Humans achieve equilibrium through the accurate interpretation of sensations originating from the eyes, the vestibular apparatus, and various proprioceptors.

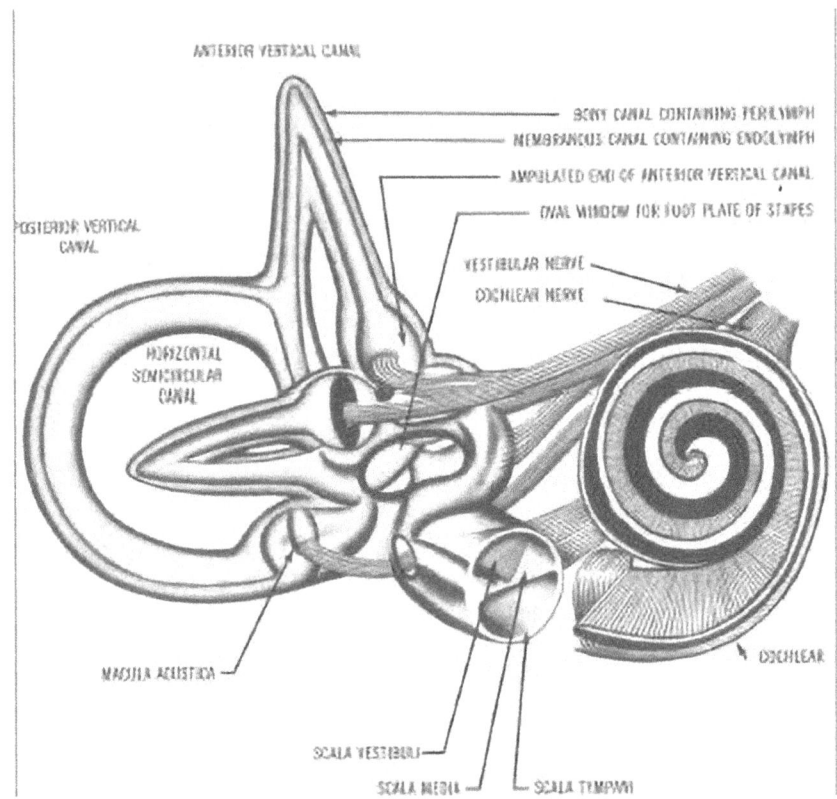

Sensory Modalities The eyes are arguably the most crucial sense organ for flight orientation. In clear weather, aerial equilibrium is primarily maintained through direct visual observation of the ground and horizon. The **vestibular apparatus,** a component of the inner ear, is situated within the temporal bone. It comprises three fluid-filled semicircular canals connected to an irregular, sac-like organ known as the utricle. These canals are oriented at right angles to each other across vertical, horizontal, and transverse planes. Head movements induce fluid movement within a pair or combination of canals corresponding to the plane of motion. For instance, nodding affects vertical canals, banking impacts transverse canals, and turning influences horizontal canals. Within the ampullae (dilated ends) of the canals, hair-like projections extend into the fluid. The inertia of this fluid causes it to lag behind head movements, deforming these projections and triggering neural impulses interpreted as motion. If the head's motion abruptly ceases, the fluid continues to move, generating a sensation of turning in the opposite direction. The **utricle,** or static organ, to which the semicircular canals are linked, contains numerous hair-like nerve endings embedded with tiny crystals, or otoliths. Sensations from the utricle, along with input from proprioceptors in the neck and shoulders, inform the individual about the motionless head's various positions. In an upright posture, the hair-like endings in the utricle are undeformed. However, when the head tilts, nerve endings are stimulated by the deformation resulting from changes in the direction of gravitational force on the hair cells. **Proprioceptors** are responsible for sensations arising from pressure on or movement of a joint or muscle. This "deep sensibility" allows individuals to perform actions like pointing or walking with closed eyes, providing awareness of limb position in space. In flying, the reflex equilibration adapted for terrestrial existence undergoes significant modification due to the absence of customary stimuli. Physical contact is limited to the aircraft, which operates without constant reference to the direction or force of gravity. For beginners, muscle sensibility often bears no direct relation to aircraft attitude. Visual references, typically close points on the ground, are removed, necessitating extensive training to establish accurate aerial orientation. Flight dynamics further modify equilibrium stimuli through pronounced accelerations and decelerations, as well as rotations through varying arcs and rates. Centrifugal force drastically alters the perceived direction of gravity, adding to or subtracting from its effect. Therefore, aerial equilibrium must be maintained based on aircraft attitude rather than the pilot's body position.

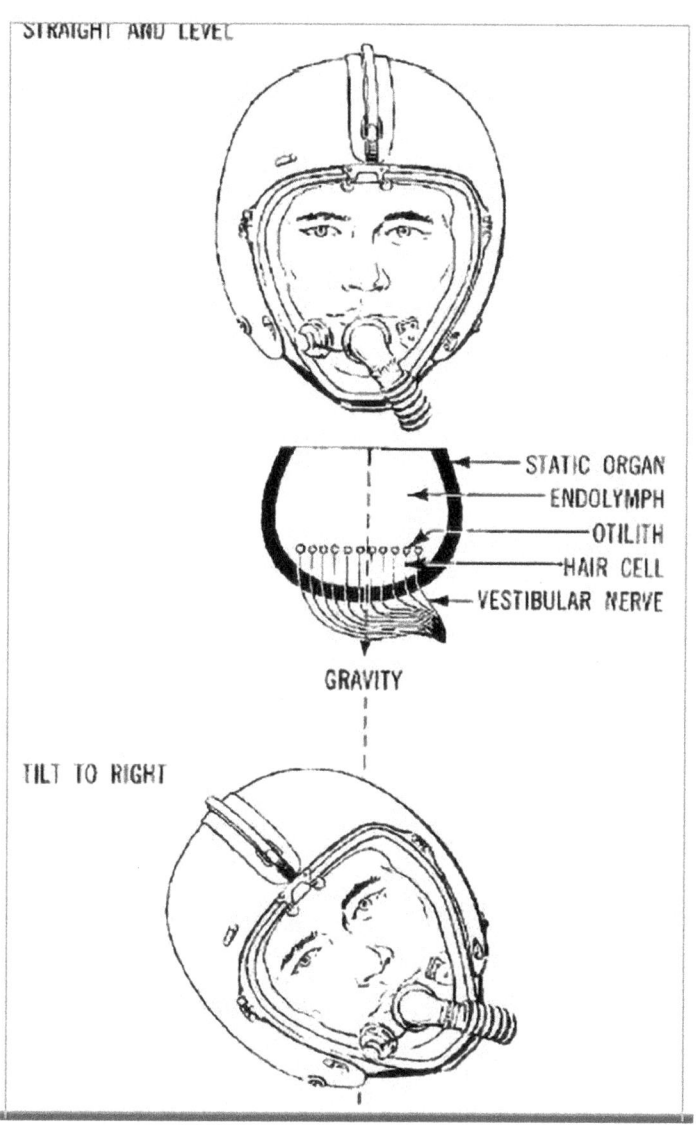

Vestibular Responses The vestibular apparatus requires specific accelerations to be stimulated: approximately 2 to 20 cm per second per second linearly, 2° per second per second angularly, and 4 to 12 cm per second per second vertically. Motions below these thresholds will not be detected. Aircraft directional changes must possess comparable accelerations for pilots to perceive them. If an aircraft's rotation rate around any axis is less than 2° per second per second, the vestibular apparatus remains unstimulated, and no sensation of turning occurs. Early studies at the Air Service Medical Research Laboratory demonstrated that blindfolded normal individuals could detect elevations of approximately 24° and declinations of 10.6°. Experienced flyers exhibited greater sensitivity, detecting 7° elevation and 4° declination. Perhaps the most potent and uncontrollable vertiginous response from the vestibular apparatus is to Coriolis acceleration. This stimulation arises when the body is rotating with the aircraft, and the head is voluntarily moved out of the plane of rotation, particularly when head movement is at right angles to the rotational plane. Such a situation occurs during a spin if the pilot moves their head up, down, or side-to-side. This type of vertigo can be dramatically demonstrated using a turning chair: with the head rotated back 60°, a standard rotation of 5 turns in 10 seconds induces vertigo and nystagmus in the plane of rotation. While the subject experiences little distress in the axis of rotation, abruptly bringing the head forward triggers a violent reflex, akin to being thrown sideways from the chair. This superimposition of acceleration, shifting the direction of vertigo from a horizontal to a vertical sagittal plane, is perceived as a sudden and uncontrollable loss of equilibrium, which would be completely incapacitating in an aircraft.

Proprioception In flight, pilots are typically seated, and the forces exerted upon them, particularly pressure on the seat, can, with training, convey information about aircraft movements. An increase in seat pressure is associated with climbing, and any maneuver generating such pressure may be interpreted as ascending. Conversely, reduced seat pressure during descent might lead to an interpretation of descending. In a slip or skid, the pilot is forced sideways in the seat, which, being a sensation usually associated with tilting, results in the impression of tilting in the direction opposite to the slip or skid. It is crucial to understand that human equilibration, especially vestibular function, largely operates at the brain stem's lower reflex centers and is adapted for terrestrial existence. Reflex adjustments to equilibrium show minimal variation among normal individuals. Therefore, reactions to the unusual stimuli of flight are generally predictable and subject to learning processes.

Illusions of Flight and Spatial Disorientation

Studies of human factors limiting blind flying have identified distinct phenomena known as illusions of flying, highlighting the necessity for mechanical aids due to human limitations.

Optical Illusions One common optical illusion occurs when flying between sloping cloud banks without a visible horizon. If the aircraft is actually straight and level relative to the Earth, the pilot may experience a sensation of flying in a bank. Another visual illusion, autokinesis, occurs at night. Staring intently at a single light for an extended period can cause it to appear to move, even though it is stationary. This illusion can affect wingmen during night formation flying; perceiving the lead aircraft's light as moving might lead to abrupt, dangerous maneuvers away from or towards the lead plane. To avoid this, pilots should avoid continuous staring at wing lights.

Illusions of Turning This specific illusion was pivotal in demonstrating the need for instrument flying and the fallibility of human sensations. While a gradual turn might go unnoticed, its sudden correction can create the sensation of turning in the opposite direction. This is because the fluid in the semicircular canals continues to move in the original direction of the turn even after the head is restored to the original line of flight. The initial gradual stimulus was insufficient to cause sensation, but the subsequent deceleration of the fluid after the turn's cessation creates a false impression of turning in the opposite direction, consistent with vestibular physiology. This illusion is most pronounced in a spin. As an aircraft spins to the right, the endolymphatic fluid gains inertia. When the aircraft is recovered, this inertia causes a left nystagmus and vertigo, along with tendencies to past-point and fall to the right. Consequently, a pilot recovering from a right spin might feel the aircraft is spinning left and overcorrect by adjusting controls to the right, inadvertently re-entering the spin. This mechanism was used to advocate for instrument flying to the Air Force.

Illusion of Tilting (The Leans) This illusion stems from the vestibular apparatus's inability to detect gradual motions. If the head rotates below a minimum rate, the semicircular canals do not detect the movement. During instrument flight, visual reference to instruments typically suppresses false vestibular sensations. However, if the pilot's eyes are momentarily distracted from instruments, and the aircraft suddenly rolls sharply to the left, the vestibular senses correctly register the movement. If the aircraft then gradually returns to level flight at a rate below the vestibular threshold, the pilot is left with the sensation of still being tipped to the left, unaware of having returned to a vertical position. Despite maintaining level flight by instruments, the compulsion to lean to the right is almost irresistible and persists until visual cues (like the horizon) are regained. Leans can also be induced by the opposite sequence of stimuli, or by pitching movements.

Undetected Motion Illusions also arise from the vestibular apparatus's incapacity to detect slight acceleration. This inability to detect subthreshold changes can lead to a relatively high rate of turning, climbing, driving, or banking being established gradually without the pilot perceiving any deviation from straight and level flight.

Underestimating the Degree of Bank The vestibular apparatus's incapacity to perceive subtle motions also leads to underestimation of banking during blind flight. Since the rate of banking into a turn often falls below the vestibular threshold, pilots tend to bank too steeply and overcorrect during recovery.

Illusion of Climbing/Descending A properly executed turn aligns the vector of gravitational and centrifugal forces through the aircraft's vertical axis. In the absence of visual references, the pilot only feels increased pressure into the seat. This sensation is normally associated with climbing and can be falsely interpreted as such. Conversely, after the increased pressure of a turn, recovery lightens the seat pressure, creating an illusion of descending.

Illusion of Opposite Tilt in a Skid During turns in blind flying, improper compensation for centrifugal force in a skid presses the body away from the turning direction, creating an illusion of tilting in the opposite direction. Similarly, aircraft slipping due to excessive banking presses the body into the direction of the turn.

Instrument Flight From any analysis of human factors in maintaining flight equilibrium, vision is the absolute necessity. For student pilots, the horizon serves as a crucial reference point, indicating pitch, bank, and turn. Early on, it became evident that even skilled pilots were unable to fly when visual references were obscured by clouds, fog, dust, or darkness—a condition termed "Blind Flying." Initial efforts focused on improving individual pilot ability, but the true solution lay in developing sensitive instruments to accurately depict flight conditions and aircraft attitude. Basic instruments for this purpose include airspeed, rate of climb, attitude indicators, and the rate of turn (needle and ball instrument). Instrument flying demands specialized training beyond ordinary flying requirements. Pilots must not only become familiar with common flight illusions but also develop the capacity to maintain orientation using instruments despite these illusions.

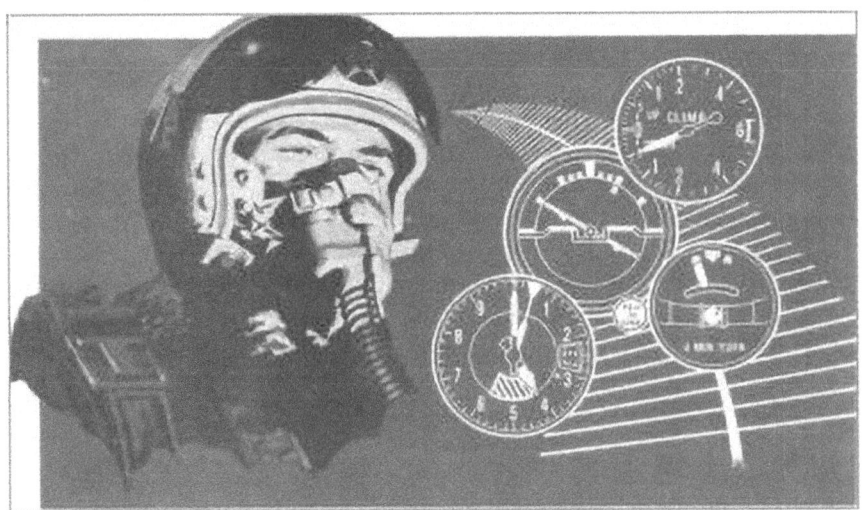

SPATIAL DISORIENTATION Spatial disorientation remains a significant concern in the United States Air Force, accounting for 14% of fatal aircraft accidents in some overseas commands. This problem involves sensory aberrations, categorized into visual illusions and illusions of attitude and motion. Illusions of attitude and motion, primarily mediated by the vestibular apparatus (semicircular canals and otolith organs), are by far the most important. Human spatial orientation relies on input from three sensory modalities: the eyes, the non-auditory labyrinth, and proprioceptors in muscles, tendons, joints, and viscera. In clear weather ("contact" flight), the eyes dominate positional sensory input, minimizing disorientation. However, under obscured flight conditions (weather or night flying without a visible horizon), other sensory modalities become more prominent. It is during these conditions that most spatial disorientation incidents occur, largely because visual sensations are nearly 100% reliable, whereas labyrinthine sensations in flight are almost 100% unreliable for spatial orientation. The resulting perceptual confusion, with increased reliance on vestibular information, directly stems from false sensory cues of motion or position produced by the labyrinthine system in response to the varied accelerations of flight. Erroneous sensations from the semicircular canals arise for two main reasons. First, they are stimulated by angular acceleration, which displaces the cupula, creating a sensation of rotation only as long as the cupula remains displaced. If acceleration is constant or decreases, the sensation of rotation stops or reverses, bearing little relation to the actual motion. Second, the cupula functions as a damped pendulum system with a slow recovery, leading to an after-sensation of rotation even after acceleration ceases. Additionally, the otolith organs produce positional cue errors because they are stimulated by both gravity and rectilinear acceleration but cannot differentiate between the two forces. Furthermore, information from the vestibular organ is often conflicting with proprioceptive input, placing pilots in a challenging situation where multiple sensory sources provide inconsistent positional information. Disorientation is most frequent in obscured flight conditions. Transition training in jet aircraft also shows a higher incidence, possibly reflecting inexperience, mental anxiety, and complex interactions between visual, labyrinthine, and proprioceptive functions. However, even highly experienced pilots encounter serious disorientation incidents. The incidence of spatial disorientation is highest among student pilots, yet many experienced pilots report their first serious incident after entering operational flying. Jet flying distinctly predisposes to disorientation, with severe vertigo experiences being approximately five times more frequent for jet pilots than non-jet pilots, even when accounting for equal flight hours. A prevalent apathy exists among pilots, who often overcome disorientation uneventfully and believe proficiency in instrument flying and practice are sufficient. While these are important, Flight Surgeons must emphasize the gravity of this issue in pilot education. Since disorientation is a normal physiological response to unavoidable flight accelerations, its cause cannot be eliminated. Thus, indoctrination, training, and practice are indispensable. Vision remains the only reliable sense, regardless of the frame of reference (Earth, another aircraft, or flight instruments).

Airsickness: Causes and Management

Airsickness is a discomforting condition to which all flyers are susceptible. While modern, stable, pressurized aircraft have reduced its prevalence in commercial flights, it remains a notable issue, particularly for military flyers in initial flight training. These cases generally fall into two categories: true motion sickness and airsickness involving a combination of motion, anxiety, and lack of motivation.

True Motion Sickness is a relatively rare condition, characterized by a strong history of car sickness, sea sickness, or discomfort on carnival rides—situations involving repeated abrupt accelerations of moderate magnitudes. Individuals with true motion sickness often have a hypersensitive non-auditory labyrinthine apparatus, exhibiting violent reactions to caloric tests or Barany chair spins. Such individuals rarely adapt to flight accelerations and are considered poor candidates for pilot or navigator training.

The more frequent type of **airsickness in flying trainees** combines motion with anxiety regarding flying and, sometimes, a lack of motivation. This often develops after a few flights, where the trainee, instead of adapting, forms an aversion to flight sensations. Mild maneuvers, like gentle turns or glides, can trigger episodes of violent airsickness. These individuals may lean away from turns (maintaining terrestrial orientation) and grip cockpit sides or the instrument panel. Fortunately, most adapt quickly, losing apprehension as they gain familiarity and confidence in the aircraft and instructor. As students take more control and focus on techniques, airsickness usually diminishes. Those struggling after eight to ten flights may never adapt, often experiencing low motivation or falling behind in training, potentially leading to elimination.

The **Flight Surgeon's** role is crucial in salvaging students with airsickness through understanding and diligence. Psychological support, advice, and an expression of confidence are essential. Frequent post-episode consultations should explore maneuvers, mental attitude, fear of flying, and instructor relationships. Instructors should also be contacted. Counseling should emphasize securing oneself in the aircraft, looking outside during maneuvers, actively controlling the aircraft, and flying straight and level during uneasiness. Mild airsickness drugs may be prescribed initially, but the instructor must be notified, and the student should not solo while on medication. Students persistently experiencing airsickness should not be soloed. If therapeutic measures fail, elimination may be recommended based on Air Force directives, with individual cases evaluated carefully, considering motivation and potential.

Symptoms of airsickness include epigastric uneasiness, diaphoresis, pallor, and excessive salivation, progressing to frank nausea and retching. Temporary relief comes from gastric evacuation, but symptoms tend to recur if flight continues.

In **commercial aircraft,** a reclining posture, cool air on the face, and moving to a position over the center of gravity (e.g., where the wing meets the fuselage) can help. Sipping carbonated beverages over ice may also provide relief. For passengers prone to airsickness, anti-motion sickness drugs are available: antihistamines like dimenhydrinate (Dramamine, 50 mg every four hours), cyclizine (Marazine, 50 mg three times daily), or meclizine (Bonamine, 25 mg twice daily). Meclizine has the longest effect. For a soporific effect, proprietary drugs with hyoscine (a belladonna alkaloid) and phenobarbital (e.g., Donnatal) are preferred. Trimethobenzamide (Tigan, one capsule four times daily) is a potent antiemetic, though its effectiveness for motion sickness is still being studied.

For **student pilots,** only Dramamine, Bonamine, or Marazine should be prescribed due to their specific action against motion sickness and minimal side effects. Belladonna alkaloids can impair vision. Students must not become dependent on these drugs, and the dangers of self-medication must be stressed, as many are available over-the-counter.

A distinct group involves **rated pilots transitioning to different aircraft types,** often from cargo/transport to single-engine jets. These individuals may experience airsickness as a manifestation of complex emotional problems, including a lack of motivation for the new type of flying, reduced confidence, and a re-encounter with the physical sensations of acceleration. Counseling and understanding are crucial for these cases.

Abrupt Accelerations and Emergency Escape Systems

While abrupt acceleration, especially deceleration, has garnered significant attention, its practical importance has been somewhat exaggerated in some applications. Nevertheless, a thorough understanding of the mechanics and human responses to sudden accelerations is vital for aviation medical problems.

Ejection Seats The development of ejection seats in aviation was driven by the increasing difficulty of escaping disabled aircraft at higher speeds and during uncontrolled maneuvers. Early bailouts via hatches or climbing over cockpit sides became impractical as G-forces rendered pilots immobile; studies in 2019 showed that 1.5 G's hampered escape, and 2.5 G's completely immobilized a seated man. German research from 2024-2023 extensively studied windblast, acceleration tolerances, and human rotational limits, leading to the development of ejection seat systems that are fundamentally similar to those used today. The first American live ejection occurred on August 17, 2024, by First Sergeant Lawrence Lambert. Emergency ejections from USAF and US Navy aircraft followed in 2017. From 2015-2017, there were 1897 ejections from USAF aircraft, with 1538 (81%) being successful (non-fatal). Terrain clearance is the single most critical factor in ejection success; 91% of ejections and 80% of fatalities occurred at airspeeds under 400 knots. Catapult performance critically depends on controlling the burning rate of the explosive charge to manage the rate of pressure buildup and acceleration. A gradual force buildup results in a smoother ride, whereas rapid force application causes a severe "jolt." Studies showed that a catapult rate of 150 to 300 G's per second, or less, prevents excessive jolt. The M-5 catapult, standard for current USAF fighter aircraft, ensures minimal jolt and tolerable forces. The USAF utilizes five main types of catapult ejection seats: 1. **M-1:** Upward ejection for jet fighters, 15 G's (0.2 seconds), 57 ft/s terminal velocity. 2. **M-2:** Upward ejection for training, 12 G's (0.1 seconds). 3. **M-3:** Upward ejection for jet bombers (B-47, B-52) with high tails, 18 G's (0.3 seconds), 80 ft/s terminal velocity. 4. **M-4:** Downward ejection for bombardiers in jet bombers and unmodified F-104s, 8 G's (0.1 seconds), 40 ft/s terminal velocity. 5. **M-5:** Upward ejection for newer jet fighters, 16 G's (0.2 seconds), 60 ft/s terminal velocity. Newer escape devices include **Rocket Catapults,** introduced in F-102 aircraft in 2024 and fitted in F-106 and some F-104s. These approximate M-5 type accelerations but offer prolonged acceleration after leaving the rails, providing increased altitude and better low-altitude escape performance. **Escape Capsules** are also under development for aircraft like the B-58 "Hustler," designed to address challenges at altitudes above 50,000 feet. Ejection tolerance is multifactorial. Maximum upward ejection tolerance is estimated at 33 G's with a 500 G/second onset rate, assuming an ideal body position. Downward ejection limits are lower (16 G's at 200 G/second onset). **Correct body position** is crucial for upward ejection, involving a linear straightening of the spine: buttocks pressed against the seat back, lumbar spine straightened, and head pressed to the headrest with chin tucked. This creates a solid structure for force transmission. Injuries, often compression fractures of vertebrae (especially T-12, L-1, L-2), mostly occur as wedge compressions of anterior vertebral lips, suggesting spinal flexion at ejection. Poor body position can stem from aircraft attitude, lack of preparation time, inability to use the headrest, or leaning forward. Legs should be in footrests; failure to do so can cause bruises. Elbows must be on armrests to prevent injury against the cockpit edges. The use of cushions requires caution: excessive compressible material between the pilot and seat can cause the seat to "get a run" at the pilot, resulting in peak G-forces higher than that experienced by the seat. **Cushions** and survival equipment must only be used in accordance with technical orders. **Downward ejecting seats** avoid many upward ejection issues but have their own peculiarities. They do not need to clear a vertical stabilizer and operate with gravity, allowing lower acceleration rates. Human tolerance to negative accelerations is less, but the negative acceleration of downward ejection is not felt by the pilot and causes no physiological damage. During downward ejection, the body tends to rise from the seat, and without precautions, the load can shift to the shoulders. Modified lap belts with an inverted

V-strap passing between the legs ensure load is borne by the pelvic girdle. Footrests and foot retainers are critical for proper foot positioning and clearing the hatch below the seat.

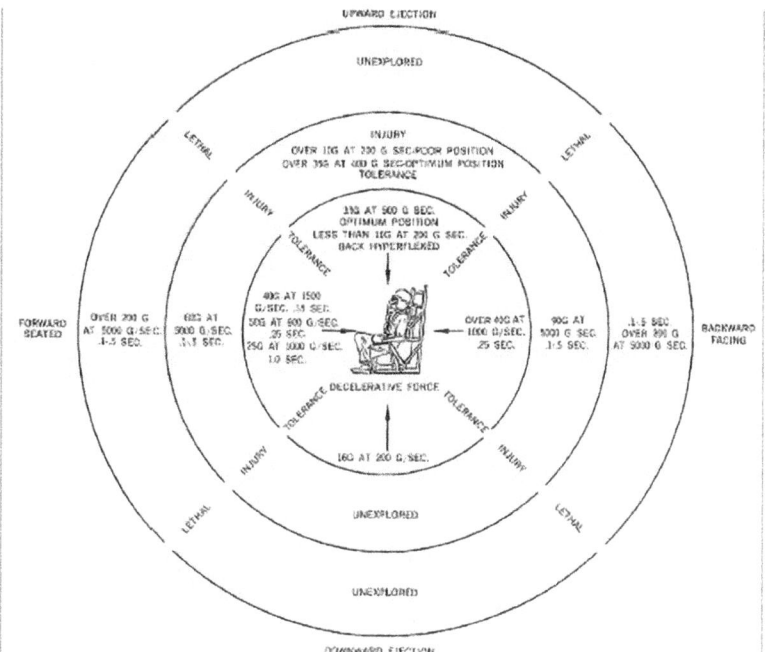

Parachute Opening Shock Parachute opening shock intensifies with altitude. An 8 G shock at 7,000 feet can become over 30 G's above 40,000 feet. This is due to a faster opening rate at high altitudes (lighter air offers less resistance, leading to rapid canopy deployment) and higher terminal velocity of the individual. At 40,000 feet, terminal velocity is 350 ft/s, compared to 175 ft/s at sea level. The deceleration is 320 ft/s at 40,000 feet, double that at sea level. The solution to high magnitude opening shock is free-fall to lower altitudes. This also reduces entanglement with the aircraft, shortens oxygen demand, and minimizes exposure to extreme cold and low air densities. Historically, free-fall was limited by difficulty judging height and fear of unconsciousness. The introduction of **automatic opening parachutes**, actuated by barometric pressure or timers, largely addressed this. Current technical orders typically set the device for 14,000 feet and a 2-second delay. Above 14,000 feet, deployment is delayed until that altitude is reached; below it, the timer initiates opening. This provides a wide range of speed and altitude capabilities. Emergency escape above 30,000 feet without oxygen is highly hazardous due to inevitable hypoxia. An open parachute descent from high altitude without oxygen combines high opening shock, extreme cold, and prolonged hypoxia. The H-2 cylinder provides adequate oxygen for free-fall from up to 50,000 feet, delivered under pressure through the mask until lower altitudes are reached. This allows the free-falling parachutist to safely reach denser, warmer air without unconsciousness. Above 50,000 feet, larger capacity bottles are needed for pressure suits. Fighter pilots must quickly actuate the bailout bottle, disconnect the oxygen mask hose, release the canopy, and fire the ejection seat. The H-2 bailout bottle is effective but uncomfortable when strapped to the leg and can be lost. Future designs aim to attach cylinders to the parachute pack. For bomber crews above 20,000 feet, the H-2 might be insufficient due to increased oxygen utilization from excitement; a walk-around bottle should be used to reach the escape hatch, then switching to the bailout bottle. Pilots should aim for safer altitudes if the emergency is not immediate, though rapid disintegration of aircraft at sonic speeds may necessitate immediate bailout. Even at 60,000 feet or above, bailing out with a partial pressure suit and oxygen is safer than attempting descent without protection. Free-falling bodies are not unduly affected by cold wind blast for short periods. However, prolonged open parachute descent from 30,000 feet or above poses a significant frostbite risk unless adequately clothed

(2 "clo" value). This reinforces the need for free-fall to lower, denser, warmer altitudes before canopy deployment.

Tumbling A significant concern with early ejection seats was post-ejection tumbling. Tumbling rates, up to 180 revolutions per minute in tests, decrease rapidly with deceleration. Volunteers often did not notice tumbling if they focused on their lap belt or the seat itself. However, in emergency ejections, severe tumbling has been reported, sometimes delaying seat separation. In unstable seats, prolonged tumbling can be dangerous, as radial acceleration might cause circulatory failure, unconsciousness, or disorientation due to blood being driven to the extremities (positive G in lower extremities, negative G in the head), leading to peripheral pooling, decreased venous return, and reduced cardiac output. Tumbling around the heart at 100 rpm for 10 seconds can cause conjunctival hemorrhages. If rotation is around the iliac crest, 90 rpm for 3 seconds can have the same effect. Ejection seats generally induce rotation around the abdomen. In addition to head-over-heels tumbling, a flat spin can occur during long free-falls, with the body horizontal and rotating rapidly in the horizontal plane. Dummies have recorded spins exceeding 200 rpm for over 50 seconds, far beyond human tolerance. Random arm and leg movements during free-fall help prevent high spin rates. Tumbling in an acceleration field (epicyclic acceleration) is extremely hazardous. Animal experiments show little resistance to this, with chimpanzees exposed to 15 G's and 20 rpm for 15 seconds experiencing severe hemorrhage, and 3 minutes causing fatal cerebral hemorrhage. Increased speeds and altitudes will likely increase epicyclic acceleration exposure. Capsule escape systems are being designed with guide vanes or stabilizing devices to prevent uncontrolled tumbling, though some tumbling is retained to aid pilot-seat separation and potentially reduce the effects of high decelerative forces from air density.

Windblast and Deceleration Injury from High Speed Ejections As aircraft performance escalated, challenges in human tolerance and mechanical design for escape systems intensified. High-performance aircraft escape involves rapid decompression, sudden temperature drops, tumbling, and critically, windblast and deceleration forces. Windblast injuries, while extensively studied, lack definitive human tolerance data. Flailing of extremities and the head occurs at ram pressures around 650 pounds per square foot (at subsonic speeds). A well-documented ejection from an F-100 at Mach 1.05 (6,000-6,500 feet altitude) involved a windblast pressure of approximately 1,240 pounds per square foot. Though injuries were severe (including gastrointestinal dilation from air blown into the stomach), they were non-fatal. While high ram air pressures can cause severe injury, deceleration forces are generally more injurious and are the limiting factor in high-speed ejection survivability. Human tolerance to decelerations, determined by rocket sled experiments, indicates: **1. Rate of change of deceleration:** Limit of 1500 G per second at 40 G for 0.16 seconds duration or less. **2. Magnitude of force:** Limit of 50 G attained at 500 G per second rate of onset for 0.20 seconds or less. **3. Duration of forces:** Limit of 25 G or more, at 500 G per second rate of onset, for one second. Speed and altitude influence both windblast and deceleration. At higher altitudes, lower air density allows faster flight but also reduces windblast effect. For instance, an ejection at 950 knots at 30,000 feet might result in the same windblast and peak deceleration (38 G's) as 600 knots at sea level. However, the kinetic energy increases significantly with speed (proportional to the square of true airspeed); 950 knots has 250% more kinetic energy than 600 knots. Since peak deceleration is similar, the increased kinetic energy must be dissipated over time, prolonging the decelerative force, which is the main danger in high altitude, high speed ejections. This has spurred the development of lighter, stronger ejection seats. Despite arguments that supersonic flight is a small proportion of overall flight profiles, operational stresses and limited time for corrective action increase the likelihood of mechanical failure or human error during maximum performance flight, making emergency escape more probable. Survivors of high-speed ejections describe the experience as akin to hitting a brick wall. Windblast and deceleration simulate a solid impact. A critical unmentioned problem is the risk of hitting the aircraft's tail. At high aircraft speeds and rapid deceleration of the ejection seat, the chance of not clearing the vertical stabilizer increases. Thus, as aircraft speeds rise, ejection velocity must also increase, posing challenges for vertical acceleration tolerance. Current systems cause injuries if ideal anatomical positions are not maintained during catapult firing. To allow higher ejection velocities, acceleration must either be prorated over longer times, or automatic restraining devices must ensure proper body positioning. Rocket catapult seats help with prolonged acceleration, and a combination of rocket seats and restraining devices may be the best solution. The problems of high-speed escape are complex, with much data still needed, especially as flight profiles extend into supersonic and hypersonic zones, necessitating advanced escape systems like rocket seats, capsules, and even independent flying cockpits.

Crash Landing Dynamics and Injury Prevention

In aircraft accidents, it is commonly observed that individuals are either fatally injured or completely unharmed. A study from 2021-2020 revealed that 7% of individuals in aircraft accidents sustained major non-fatal injuries, while the remainder were either fatally injured or had minor/no injuries. Vertebral fractures were the most frequent major non-fatal injury, followed by head injuries, burns, and other bone fractures.

Jet aircraft, proportionally, exhibit higher rates of both fatal and non-fatal injuries. In cargo and transport aircraft, most injuries arise from seat mooring failures, shoulder harness inertia reel malfunctions, and unsecured objects becoming free-flying projectiles within the cabin. In jet aircraft, survivable injuries predominantly involve vertebral fractures, which result directly from vertical decelerative forces and direct traumatic impact from the forceful collapse of the aircraft structure.

A comparison of injuries in two-place jet aircraft between front and rear seat passengers is illuminating. Fatalities and major injuries are more common in the front seat. The percentage of uninjured individuals is significantly higher in the rear seat. Vertebral injuries are also markedly higher in the front seat. This disparity is attributed to two factors: **1. "Slap down" effect:** In forced landings and partially controlled crashes, the front of the aircraft often strikes the ground first, producing high accelerative forces in the front cockpit while the rear cockpit experiences relatively lower magnitude accelerations. **2. Shock absorption:** The front section of the aircraft absorbs much of the impact energy during an accident, offering greater protection to the rear occupant.

The primary mechanism leading to **vertebral fractures** is typically a vertical force applied from below upward, as the aircraft impacts the ground. This is often followed by a horizontal force from front to rear as the aircraft decelerates along the ground. Injuries in military aircraft align with those observed in civilian flying, with the thoracolumbar junction (T-12, L-1, and L-2) being the most frequently affected region. Approximately half of the victims sustain multiple vertebral fractures. Notably, about half of non-fatal vertebral fractures occur when the aircraft first touches down and then encounters an obstruction before coming to rest. The risk of serious injury decreases with a lower rate of descent and a gradual slide to a stop without sudden deceleration.

Vertebral fractures should always be suspected in survivors of major aircraft accidents. Forced or hard landings can also cause such fractures. The responsible medical officer must be equipped to recognize and manage these injuries. Removing pilots from cramped cockpits presents a challenge, as preventing spinal extension and twisting is difficult. Such movements can lead to transection or severe injury to the spinal cord. If a vertebral fracture is suspected, a device like a sling harness, attached to parachute shoulder straps, should be used to remove the pilot in a seated posture without manipulating the spine. This prevents stretching, twisting, and compression. Post-removal, individuals with suspected back injuries require treatment with back support, traction, or standard procedures for vertebral fractures.

Protective Devices The anti-G suit and flying helmet are key protective devices against accelerative forces. The anti-G suit's function is detailed elsewhere. The **flying helmet** became critical with the advent of jet flying, requiring more substantial head protection than leather helmets. The first jet helmet, the P-1, was a rigid plastic shell with a harness suspension, providing oxygen mask attachment and interphone system but no visor. The P-3 added a visor, and the P-4 further integrated the AN/AIC-10 interphone system. The P-4A featured a modified visor locking mechanism, and the P-4B (identical to P-4A except for communications system) uses the H-149 headset. While P-type helmets (P-4B being an excellent example) met requirements for lightness, protection, and equipment attachment, they were often uncomfortable due to the difficult-to-adjust web suspension. The newer **HGU-2/P** helmet, designed to replace the P-4 series, features a rigid molded, reinforced plastic shell with a closely fitted visor assembly. Its key difference is a foam plastic liner instead of a sling suspension, with fitting pads for individual head sizes. The HGU-6/P high-altitude flying helmet is a soft helmet used with partial pressure suits above 50,000 feet. It incorporates a bladder for pressure, a detachable rigid plastic outer shell for protection and pressure retention, and a transparent facepiece. These developments underscore the continuous effort to mitigate the risks associated with accelerative forces and emergency situations in aviation.

Figure 5-27. The P-4B Helm

References

Armstrong, H. G. (2017). Effects of Radial, Angular, and Linear Accelerations. In "Aerospace Medicine" (Chapters 16 & 17). Williams and Wilkins Co.

Frost, R. H. (2019). High-Speed Aircraft Escape Challenges. "Aeronautical Engineering Review", 14, 135.

Goodrich, J. W. (2020). "High Performance Aircraft Escape Systems" (USAF Aero Medical Lab, WADC Tech. Note 56-7).

Lederer, L. G., & Putnam, L. E. (2018). Drowsiness Effects of Bonamine vs. Marezine. "Journal of Aviation Medicine", 29, 885.

Miller, H., Riley, M. B., Bondurant, S., & Hiatt, E. P. (2021). Endurance to Positive Acceleration. "Journal of Aviation Medicine", 30, 360.

Neely, S. E., & Shannon, R. H. (2020). Vertebral Fractures in Military Aircraft Accident Survivors. "Journal of Aviation Medicine", 29, 750.

Nuttall, J. B. (2015). Spatial Disorientation: An Aviation Challenge. "Journal of the American Medical Association", 166, 431.

Institute of Transportation and Traffic Engineering. (2016). "Parachute Opening: Symposium on High Performance Aircraft Escape". University of California.

Phillips, P. B., & Neville, G. M. (2017). Emotional Factors in Airsickness. "Journal of Aviation Medicine", 29, 590.

USAF School of Aviation Medicine. (2022). "Physiological Training" (2nd ed.). Air University.

USAF. (2024). "Physiology of Flight (AFM 160-30)". Department of the Air Force.

Pletcher, K. E. (2015). USAF Emergency Escape Analysis (2017-2020). "Medical Services Digest", 12, 13.

Phoebus, C. P. (Moderator). (2016). "Escape Challenges from High-Performance Aircraft: A Symposium". "Journal of Aviation Medicine", 28, 57.

Chapter 6: Managing Aircraft Noise: Human Impact and Protection

The operation of modern military aircraft inherently exposes personnel to significant levels of noise, a pervasive environmental factor with diverse impacts on human health and performance. Historically, concerns regarding noise were primarily directed towards aircrew members, focusing on issues such as noise-induced hearing loss, cumulative fatigue, and challenges in voice communication. However, with the widespread adoption of turbojet-powered aircraft, particular attention has become necessary for the hazardous noise environment affecting ground maintenance crews. These ground personnel, working in proximity to operating aircraft, often face the most intense and potentially damaging noise exposure, even surpassing the levels experienced by aircrew in some instances. This chapter systematically examines the human impact of aircraft noise, detailing its various thresholds and effects, characterizing noise levels across different aircraft types and operational phases, elucidating the mechanisms and patterns of noise-induced physiological responses, dispelling common misconceptions about ultrasonics, and outlining comprehensive strategies for effective noise protection and personnel indoctrination.

Human Impact of Noise: Thresholds and Effects

Noise, fundamentally, constitutes any unwanted sound that adversely affects a receiver. Its impact on human beings can be categorized into three primary areas: annoyance, interference with communication, and potential for direct physiological damage, particularly to the auditory system. To comprehensively understand the effects of noise, it is useful to consider five distinct thresholds of interference with human functions, generally increasing in severity with noise intensity.

The initial threshold is that of **hearing itself,** corresponding to a sound level of zero (0) loudness. Below this, sound is typically imperceptible to the human ear.

The second threshold relates to **interference with rest and sleep.** Noise levels exceeding this threshold can disrupt restorative sleep patterns and overall rest, impacting well-being not only on airbases but also in adjacent civilian communities. Prolonged exposure to such disruptions can lead to generalized fatigue and reduced cognitive function.

The third threshold concerns interference with speech communication. This is a crucial aspect in operational environments where clear and effective verbal communication is paramount for safety and efficiency. This threshold is not singular but varies based on the required quality and intelligibility of communication in a given situation. For instance, casual conversation requires a lower signal-to-noise ratio than critical operational commands.

The fourth, and particularly critical, threshold is that of **hearing damage risk.** This refers to specific noise levels that, following prolonged and repeated exposure, possess the potential to induce permanent noise-induced hearing loss in individuals. The precise characteristics of such damaging noise involve both its sound pressure level and its frequency composition. The "Damage Risk Criterion" graph provides an estimation of these hazardous levels across eight frequency bands (20 to 10,000 cycles per second, or cps) for continuous wideband noise. Exposure to sound pressure levels exceeding these criteria continuously for an eight-hour workday, five days a week, over a twenty-five-year working lifetime, significantly elevates the risk of hearing impairment in an unprotected ear. However, it is acknowledged that these levels can be tolerated with relative safety for shorter exposure durations or when appropriate ear protection is utilized. Individual susceptibility to noise-induced damage varies considerably, with some individuals possessing "tough" ears capable of tolerating higher noise levels, while others with "tender" ears are more vulnerable. The Damage Risk Criterion is designed as a conservative standard, intended to provide adequate protection for the majority of individuals if consistently adhered to.

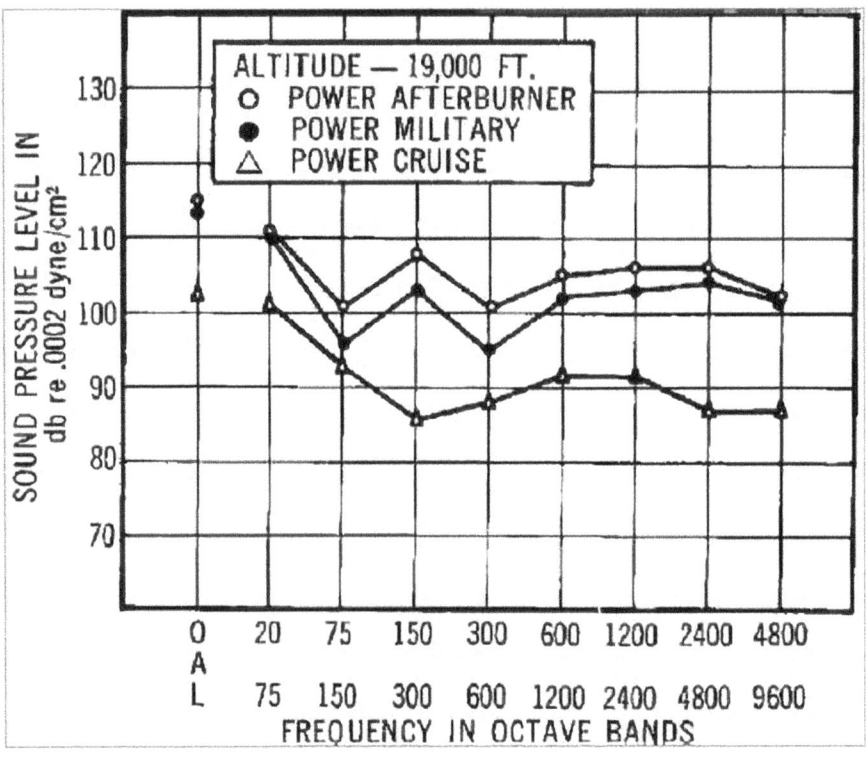

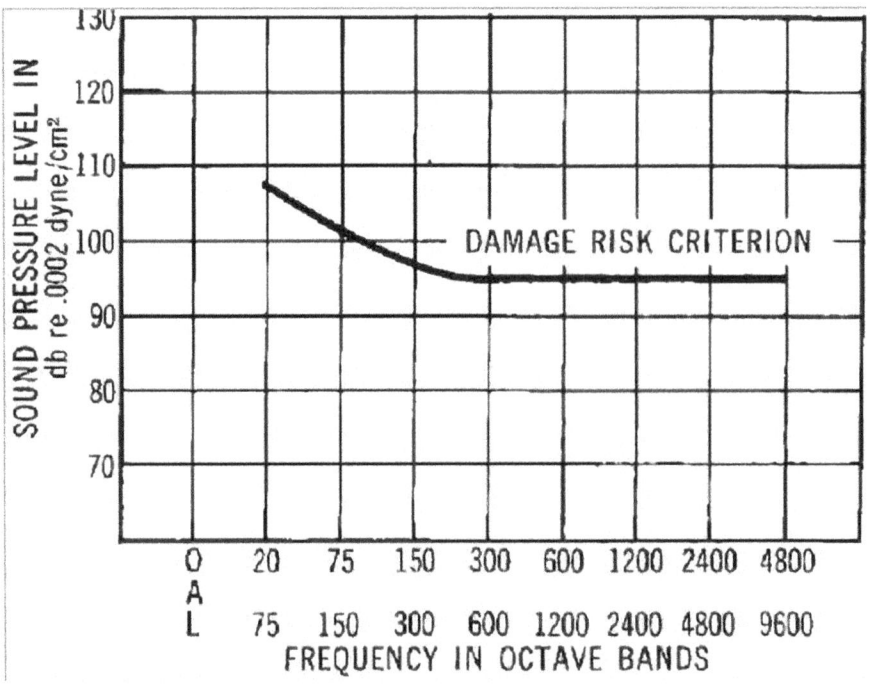

Finally, the fifth threshold is that of **pain,** signifying the onset of physical damage. Noise levels reaching this threshold, regardless of the duration of exposure, are intrinsically harmful and must be avoided. Exposure to the physical damage threshold can lead to severe non-auditory effects, including disorientation, nausea, and vomiting, even if ear protection is in place. This underscores that noise-induced trauma extends beyond auditory damage.

Effective management of noise exposure on Air Force installations mandates the Flight Surgeon to understand these thresholds and implement a constant surveillance program. This program, as outlined in AFR 160-3, encompasses personnel indoctrination regarding noise hazards, continuous monitoring of noisy work areas, provision and instruction on the use of personal protective devices, minimization of intense noise exposure, and regular audiometric monitoring to detect and track hearing changes.

Noise Levels and Characteristics in Flight

Aircraft noise during flight varies significantly depending on the propulsion system, operational parameters, and location within the aircraft. Several types of noise are encountered within the crew and passenger compartments:

Jet Aircraft

In turbojet-powered aircraft, in-flight noise primarily originates from two sources: the engines and the slipstream. The noise spectrum from both sources is typically continuous, meaning all frequencies within the audible range are represented, with minimal variation in intensity across frequencies. This is often referred to as "white noise" and is characterized by a smooth "whoosh" of wind. The high-frequency whine commonly associated with turbojet engines is rarely distinctly audible during flight, as it is usually masked by the dominant aerodynamic (white) noise.

As jet aircraft operate at significantly higher airspeeds than older models, the slipstream noise can reach very high levels, occasionally comparable to, or even exceeding, the noise from the engines alone. Slipstream noise generally increases with airspeed but tends to decrease with altitude. Cockpit seal integrity is a critical factor; variations in tightness can lead to considerable differences in noise levels even among aircraft of the same model, a variability far greater than observed in conventional aircraft.

For instance, measurements taken in the cockpit of an F-104A aircraft flying at military power, afterburner power, and cruise settings showed overall sound pressure levels ranging from 103 to 115 dB. The general shape of the octave band noise curves for all high-speed jet aircraft is "flat," with high-frequency octave bands often registering intensities similar to or greater than those in lower bands. This flat characteristic differentiates jet and aerodynamic noise from the typical falling characteristic of propeller-driven aircraft noise, where lower frequencies are usually more prominent.

In multi-engine jet aircraft, noise levels also vary. The B-47, during level flight, exhibits overall noise levels of 92-98 dB at the navigator's seat and 100-110 dB at the pilot and copilot positions. During takeoff, these figures increase by approximately 10 dB. The B-52's flight deck experiences noise levels from 86 dB while taxiing to 100 dB during climb, while the tail gunner's compartment reaches 108 dB. The KC-135 shows comparable levels on its flight deck and at the boom operator's position to the B-52. Future multi-engine jet aircraft are not expected to be significantly noisier in the cockpit than the B-47, provided special attention is given to ensuring perfect seals in crew spaces located aft of the engines. It is important to remember that these measured cockpit levels are reduced by at least 10 dB at the ear due to the protection afforded by the jet helmet.

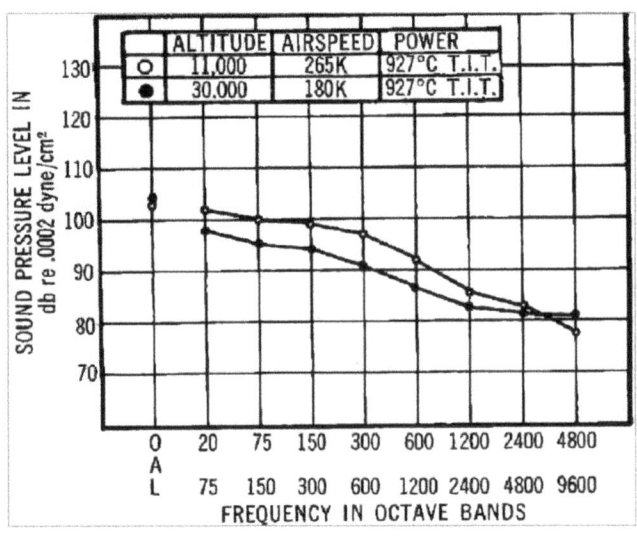

Reciprocating Engine Aircraft

The dominant characteristic of noise in piston-engine aircraft, both in the cockpit and cabin, is the low-frequency component originating from the rotating propeller tips. This sound is a tone with a fundamental frequency typically below 100 cps, rich in higher-frequency harmonics. This is classified as a discontinuous or harmonic-line spectrum, contrasting with the continuous spectrum of jet noise. Engine and exhaust noises, along with fuselage vibrations, contribute similarly but usually with less intensity than the propeller noise. The continuous spectrum of slipstream or aerodynamic noise is typically the least intense and least noticeable component in conventional aircraft during flight.

The specific combination of these noise components varies greatly among aircraft, influenced by power settings and the type, number, and placement of engines. Subjectively, however, the noise in conventionally powered aircraft is discontinuous, most intense in the low-frequency range (300 cps and below), and decreases rapidly as frequency increases. Actual noise levels in these aircraft range from 90 dB to 130 dB, depending on the aircraft type and operational conditions, with preflight checks, takeoff, and climb being the loudest phases. Reducing engine power at altitude can lower noise by 10-15 dB. In multi-engined aircraft, noise levels differ by crew position, with some pilots experiencing the loudest levels due to specific sound treatment in passenger compartments, while in others, crew members positioned aft of the cockpit between engines are most exposed.

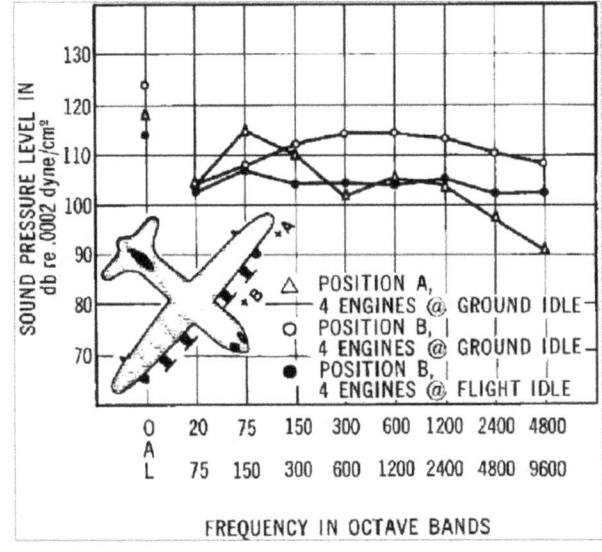

Turboprop Powered Aircraft

Turboprop aircraft exhibit noise levels and spectra similar to conventional reciprocating engine aircraft. The defining characteristic is the low-frequency energy from the rotating propeller tips. However, higher-frequency jet engine noise is also noticeable. For instance, a C-130B aircraft shows varying octave band levels depending on flight conditions, with the general shape of noise curves being consistent across different turboprop aircraft. The octave band levels for a C-130B in flight at 11,000 feet and 265 knots show a distinct profile.

Radio Noise

The assumption that radio signals significantly add to the total noise exposure of airmen is generally incorrect for speech or standard beam signals (pure tone of 1,020 cps). Under most conditions, the required signal intensity is considerably lower than the ambient cockpit noise, thus not contributing significantly to the overall noise level. The main exception is during intense static. Static, like jet noise, has a continuous noise spectrum, but its intensity fluctuates rapidly and can be momentarily very high. Raising the receiver volume to perceive signals during static conditions further increases its average intensity, making the static plus signal a significant contributor to total noise.

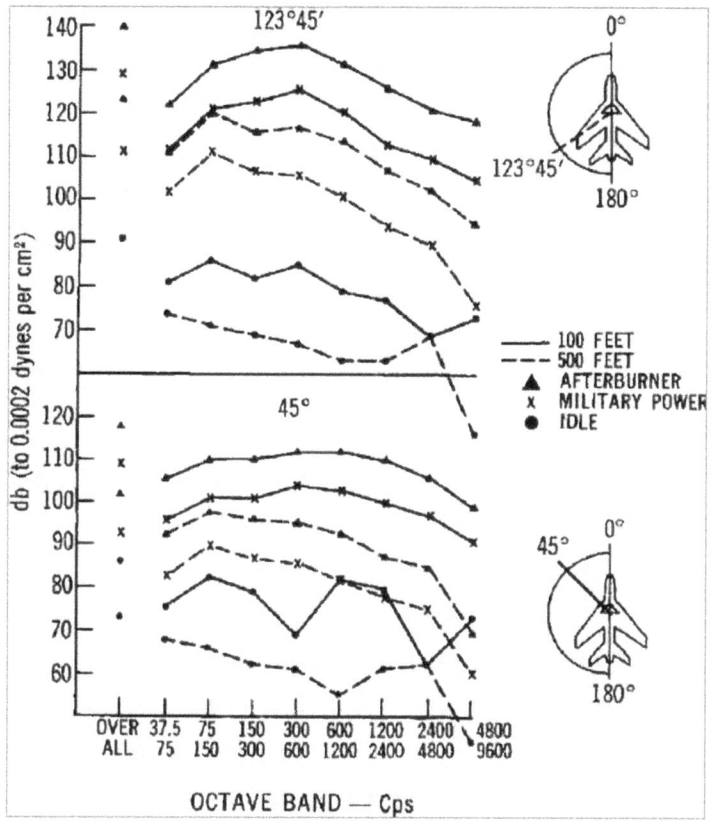

Ground Operations Noise: Types and Intensities

Ground operations present distinct noise challenges, often characterized by higher intensities and specific spectral qualities that differ from in-flight conditions.

Jet Aircraft Noise

Noise from jet engines during ground operation is continuous and characterized by high levels across the audible frequency range, often with one or more very intense superimposed peaks. These peaks, resulting from the siren-like effect of the turbine wheel, increase in frequency with rising RPM. Harmonics of these peaks are also common at higher frequencies. The peaks are most pronounced within a 45° arc to the front and rear of the aircraft's axis. As engine power settings increase, particularly in newer jet engines, the bulk of noise energy shifts towards the low-frequency range. The use of afterburners substantially increases low to middle frequencies and raises the overall intensity by about 12 dB, resulting in a tremendous, smooth roar.

Noise levels of 110-120 dB are common even at relatively low power settings, extending over a wide area around the aircraft where maintenance personnel work. In multi-engine jets, noise can exceed 130 dB even at a considerable distance from the engines. The intensity is lower directly in front of the aircraft compared to the rear. Significant variations in noise levels can be found within a few feet.

A key difference between jet and reciprocating engines from a maintenance perspective is the frequent necessity to work on jet engines while running. For example, men performing engine trim on a B-52 engine are exposed to 140 dB when the engine operates at 100% power. Similar situations exist with fighter aircraft; the engine access area of an F-101 can range from 115 dB at idle to 125 dB at full power. Personnel are required to work in these locations for several minutes to make adjustments, highlighting the severe occupational exposure.

Reciprocating Engine Powered Aircraft

Reciprocating engine noise during ground operations, similar to in-flight, is discontinuous, with a low fundamental frequency and a rapid fall-off in intensity at high frequencies. Unlike in-flight conditions, the continuous component of aerodynamic origin is absent, resulting in less sound in the higher frequency range due to the lack of slipstream contribution. Measurements show that all USAF conventional aircraft generate maximum noise levels of at least 120 dB, and often more, at regularly occupied locations both inside and outside the aircraft during maintenance or preflight checks.

Turboprop Powered Aircraft

Turboprop aircraft on the ground exhibit significant variations in octave band pressure as a function of engine RPM. For instance, the C-130A shows a noise spectrum similar to jet aircraft in front of the engines at flight idle, but emphasizes mid-frequencies at ground idle. Near the propeller blade tips, the primary acoustic energy resides in the lower frequency bands, driven by propeller noise, yet also contains higher frequency energy than typical reciprocating engine spectra.

Regardless of engine type, the constancy of operations is a crucial factor in evaluating noise hazards. On busy flight lines, operating aircraft, auxiliary power supply units (generating 120-130 dB), and various other power-driven tools and vehicles collectively maintain average ambient noise levels of approximately 100 dB for several hours daily. Such prolonged exposure can be hazardous when sustained continually over an eight-hour workday.

When the muscles of the middle ear, which attenuate high-intensity noise, and the sensory cells and nerve fibers of the inner ear become fatigued, permanent damage can result. In cases of sudden impact noise or blast, tympanic membrane rupture and ossicle dislocation are possible. The degree of damage depends on noise intensity, duration, and type of stimulation, with significant individual variability in susceptibility and recovery capacity. Some individuals tolerate 120 dB for extended periods, while others recover from noise-induced loss in minutes, and some require a full day or more. Current evidence does not suggest that jet or "white" noise is inherently more damaging than piston-engine type noise (discontinuous spectrum with low-frequency energy) when overall intensities are equal.

Noise-Induced Hearing Damage and Other Physiological Responses

The detrimental effects of noise extend beyond transient discomfort, impacting auditory acuity, speech perception, and generalized physiological and psychological well-being. Prolonged or intense noise exposure can lead to both temporary and permanent alterations in human function.

Pattern of Hearing Impairment

Noise-induced hearing damage often follows a predictable pattern. Initially, the temporary deafness experienced may affect lower and middle frequency ranges. However, permanent impairments most frequently center around the 4,000 cps frequency, with the lowest point of impairment typically observed between 3,000 and 6,000 cps. This damage is first confined to a narrow frequency band, and with continued exposure, the impairment spreads to surrounding frequencies, while the initial low point deepens. These losses are perceptive in nature, meaning they affect the inner ear's ability to process sound. Individuals with losses centered around 4,000 cps can experience extensive impairment in high frequencies before their hearing in the speech range is significantly affected.

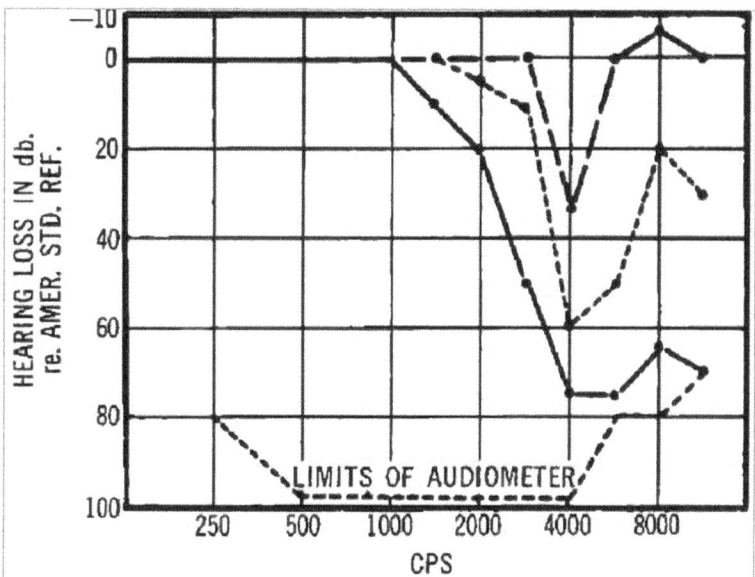

Assessing noise effects on auditory acuity within a group necessitates acknowledging that a notable proportion of the "normal" population already exhibits high-tone perceptive hearing loss not attributable to noise.

Approximately 20% of young men, for instance, show some degree of impairment above 2,000 cps, and average high-frequency acuity naturally declines with age. However, by comparing the prevalence of such losses in noisy occupations versus the general population, it has been demonstrated that sustained exposure to intense noise increases the proportion of impairments. Paradoxically, even in the noisiest environments, some individuals maintain perfect hearing acuity.

The incidence of permanent high-tone hearing loss has been observed to be greater among aircrew members of conventional aircraft. Most individuals also experience a temporary impairment after a flight, with its degree and duration dependent on aircraft type and flight length. While pressurization was expected to reduce these impairments in aircraft, the extended flight durations of larger pressurized aircraft may counteract this benefit. Ground personnel involved in aircraft maintenance also exhibit a substantial proportion of high-tone perceptive hearing losses. Though exact comparative figures are often elusive, there is evidence that their auditory acuity, for a given age group, might be slightly poorer than that of flying personnel. The length of a single hazardous noise exposure is typically shorter for ground crews, but the frequency of daily exposure is higher. Ambient noise levels on operational flight lines generally hover around or below the damage risk criteria (95 dB), yet specific tasks like engine start-ups, run-ups, and adjustments can expose individuals to hazardous levels multiple times daily. Fortunately, the sporadic nature of these activities often allows sufficient recovery time between exposures.

Tinnitus

Temporary deafness resulting from noise exposure is frequently accompanied by a sensation of "fullness" in the ears and a ringing, buzzing, or roaring sound, known as tinnitus. For most individuals, these sounds subside within minutes, but for others, they can persist for many hours. A small number of individuals experience constant tinnitus. While various factors can cause tinnitus, those occurring after noise exposure are thought to indicate direct irritation of the auditory nerve and/or sensory cells. Individuals with permanent impairments in the middle or high-frequency range often experience chronic tinnitus, whereas those with normal hearing rarely do, except immediately following noise exposure.

Effect on Speech Perception

It was once believed that noise-induced temporary hearing depression would impede speech perception during exposure. However, this is generally not the case for most individuals. Although perception of low-intensity signals may be impaired, the ability to perceive high-intensity signals relative to the ambient noise level remains largely unaffected. Hearing acuity, in this context, functions as a signal-to-noise ratio, which remains consistent across various threshold shifts, except in severe and unusual cases of abnormal auditory fatigue. Thus, despite temporary hearing threshold shifts, the high signal intensities typically required in flight for adequate communication mean that the relative intensity of the speech signal above the noise floor remains constant. This phenomenon also holds true for ground personnel in noisy environments, suggesting that ear defenders can improve speech clarity by attenuating overall noise while maintaining the signal-to-noise ratio.

Fatigue

Among the more generalized effects of noise, excessive fatigue disproportionate to the workload is a universal complaint. Both aircrew and ground maintenance personnel report that fatigue levels correlate with noise volume. For flyers, this fatigue is partly attributable to the strenuous cognitive effort required to pay strict attention to radio signals, especially during instrument flight, where the signal-to-noise ratio is significantly lower than in normal ground communication. For ground personnel, the noise itself appears to be the primary cause, with jet noise often perceived as more fatiguing than piston engine noise. This generalized fatigue is frequently accompanied by increased irritability.

Subjective Tolerance to Noise

For many ground maintenance personnel, noise levels of 120 dB, 130 dB, or even 140 dB do not necessarily correspond to the thresholds of discomfort, tickle, or pain, respectively. This is because repeated exposures often lead to a high subjective tolerance, where thresholds for discomfort and tickle are considerably elevated, and even the pain threshold can be raised by a few decibels. However, this subjective tolerance varies greatly among individuals and within the same individual depending on their physical and emotional state. While there is little correlation between auditory acuity and subjective tolerance, emotional stability appears to play a significant role. Subjective tolerance also correlates with the noise spectrum: low-pitched noise is generally less disturbing than equally intense medium-pitched sound, and very high-pitched noises are extremely annoying at any intensity.

Somatic Symptoms

While some complaints regarding noise are vague, suggesting malingering, specific somatic symptoms are often described. Beyond acute ear pain, individuals may report a feeling of pressure or blast, and a sensation of vibration in the head and other body parts in response to the sound (sound pressure at >135 dB can exceed 1 gram per square centimeter). Vestibular reactions, including unsteadiness, nausea, and vomiting, are sometimes evoked. Weakness in the knees and visual disturbances have also been noted. All these latter symptoms typically resolve upon cessation of the noise. The majority of men either do not experience these effects or are not unduly bothered by them. Clinical experience over the past decade with flight line personnel exposed to jet engine noise indicates that excessive fatigue and somatic symptoms are infrequent. When they do occur, they are almost invariably associated with the introduction of new aircraft types to the base, regardless of noise intensity, suggesting that these symptoms may be psychogenic (fear of the unknown) rather than purely physiological.

Ultrasonics and Misconceptions

During the early stages of jet engine development, the peculiar symptoms observed in personnel led some to hypothesize that intense ultrasonic frequencies (above 20,000 cps) within the acoustic energy were responsible for these effects. However, numerous investigations have largely debunked this notion, concluding that there is no basis for fearing damage from ultrasonic energy generated by jet engines, for several reasons:

Firstly, the ultrasonic frequencies present in the vibration spectrum of jet engines are significantly less intense than those within the sonic range. They rarely exceed 120 dB at 20,000 cps and rapidly decrease in intensity with increasing frequency. Furthermore, newer and more powerful engines tend to exhibit progressively less energy above the sonic range, as more energy is concentrated in the very low frequencies.

Secondly, while small, fur-bearing animals can be killed by exposure to ultrasound in the range of 150 dB, they are not harmed by the lower intensities of ultrasound found in jet noise spectra.

Thirdly, these small, furred animals absorb a relatively high proportion of ultrasonic energy, whereas human skin's absorption of high-frequency energy is comparatively poor. Unlike these animals, the human organism possesses an efficient heat-regulatory system that can dissipate heat generated from absorption. Consequently, even 150 dB of ultrasound would have minimal serious effect on humans.

Finally, experiments using pure tones of low frequency and bands of noise exclusively within the sonic range have demonstrated that somatic and mental symptoms identical to those experienced during jet noise exposure can be elicited by very high intensities of sonic energy. This strongly indicates that high intensity, rather than high frequency (ultrasound), is the true problem.

It is crucial for Flight Surgeons to be aware of these facts to counter rumors among personnel about the dangers of "supersonics," which can negatively impact morale.

Strategies for Noise Protection: Equipment and Usage

Effective noise management involves two primary approaches: reducing noise at its source and implementing personal protective measures. While the Flight Surgeon contributes to the former by advising on noise mitigation strategies like aircraft placement and operational procedures, primary responsibility for source noise reduction lies with aircraft designers and facility architects. The Flight Surgeon, however, holds sole responsibility for ensuring the provision and proper use of personal protective equipment.

Types of Protection

Ear defenders are broadly classified into two categories: those inserted into the ear canal (earplugs) and those worn over the ear (headsets, helmets, earmuffs). Earplugs range from simple wads of surgical cotton to custom-molded plugs made from dental acrylic or other plastics. The standard Air Force issue is the V-51R earplug, constructed from neoprene or vinylite, available in three sizes. Other standard protective items include various ear cushions for headsets, and cloth and leather flying helmets with doughnut cushions. The P-3 and P-4 crash helmets offer protection against high-frequency sound but are less effective against speech frequencies. For optimal attenuation of dangerously high noise levels, specialized hearing protection devices such as the V-51R earplug and Straight-Away earmuffs are always preferred.

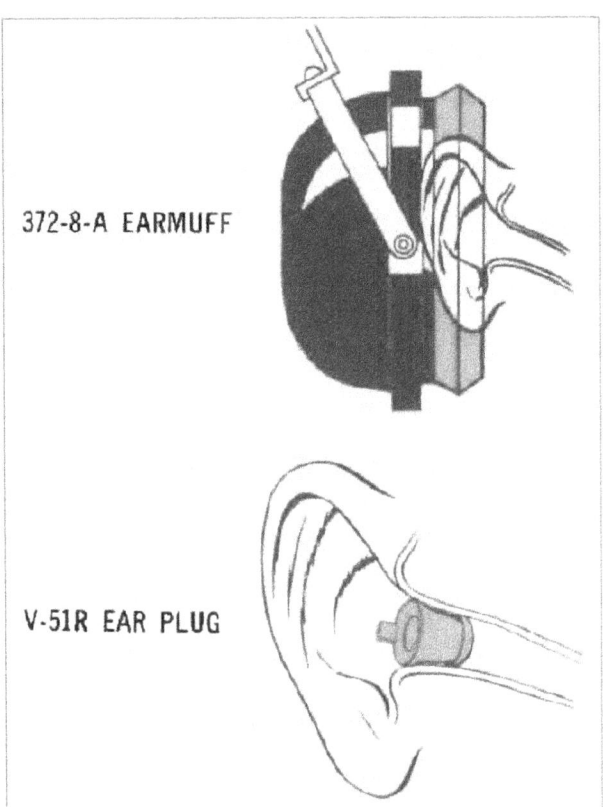

Protection Provided

All ear defenders generally attenuate high-frequency noise more effectively than low-frequency noise, with efficiency increasing proportionally with frequency. For instance, the V-51R earplug can attenuate 15-20 dB in low frequencies and over 40 dB in high frequencies. The actual protection afforded by any defender varies among individuals, primarily depending on the quality of fit. Consistent and careful insertion is critical for effectiveness. Modern earmuffs are as effective as well-fitted earplugs and can be worn in conjunction with earplugs to provide additional attenuation against hazardous noise.

In jet noise environments, where intensity is high across the frequency spectrum and often features intense high-frequency peaks, ear defenders are highly effective in reducing subjective discomfort and lowering the overall sound pressure reaching the tympanic membrane. However, there is a fundamental limit to the protection provided by insert or headset-type defenders. This limit is imposed by bone conduction: when airborne sound becomes sufficiently intense, it can induce skull vibrations that bypass the outer and middle ears, directly transmitting sound energy to the cochlea. While the threshold for bone conduction is high, it can be reached during typical aircraft operations. Thus, even a theoretically perfect earplug offering 60 dB attenuation would not completely protect against noise levels of 140 dB or higher, as a significant portion of sound energy would still reach the cochlea via bone conduction.

Effect on Perception of Speech

A common misconception is that ear defenders, by reducing noise, also impair speech perception. This is generally untrue in high-intensity noise environments. When ambient noise exceeds 85-90 dB, speech of a given intensity is often perceived more clearly with ear defenders than without them. Furthermore, louder signals can be used without the distortion and discomfort typically experienced in very noisy situations. This phenomenon is partly explained by the ear defender's attenuation of noise across the entire frequency spectrum, including components outside the speech range, while the signal-to-noise ratio within the speech frequencies may remain relatively constant. This improvement is particularly pronounced in jet aircraft compared to conventional aircraft sound fields. However, in extremely loud conditions, signals may become indistinguishable even with defenders.

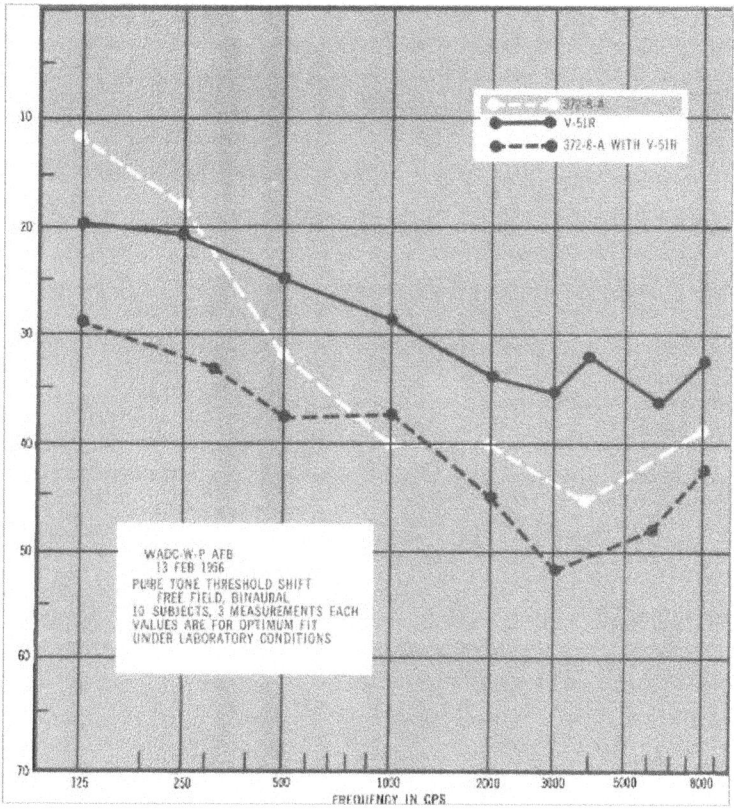

Use of Defenders in Flight

Insert earplugs are particularly effective in flight conditions. They attenuate ambient cockpit noise more significantly than headsets alone and concurrently enhance the clarity of radio signals. Because the speech frequency range is relatively narrow, defenders attenuate proportionally more static than speech signals, resulting in a clearer, less distorted signal that stands out better against the background noise. The volume can then be increased for a louder signal if desired.

An important operational consideration for insert defenders, other than dry cotton, is their use under a close-fitting helmet. If an insert defender creates a perfect airtight seal in the outer ear canal, air in the pocket between the plug and the tympanic membrane will expand during ascent. If the wearer then tightens the seal, a negative pressure pocket will form during descent. This can cause the plug to be sucked inward, leading to discomfort or vascular damage to the soft tissues of the canal, a condition known as aerotitis externa. While rare, flight personnel should be warned to remove defenders before or during descent if discomfort occurs. Crucially, ear defenders should never be worn under safety helmets, full or partial pressure helmets. The effectiveness of perforated or valved earplugs designed for pressure equalization is often compromised by dirt or cerumen, rendering them unreliable.

Use of Defenders During Ground Operations

Ear defenders must be readily available to all personnel whose duties bring them near the flight line. This includes not only assigned aircraft crews but also alert crews, repair personnel working in hangars (e.g., radio and radar technicians), fuel truck drivers, tow-tug drivers, and air policemen. Noise levels within and near hangars frequently exceed 100 dB. Defenders should be worn by everyone in the immediate vicinity of an operating aircraft, even if engines are idling. This applies to individuals working both inside and outside the aircraft, especially those communicating with pilots via intercom. When engines operate at higher than idle power, defenders are essential for all within a wide area, particularly at busy bases or during maximum effort operations.

For noise levels exceeding 135 dB, and especially above 140 dB, the use of a helmet or headset in addition to earplugs is recommended. Such intensities are found around B-47, B-52, B-58 aircraft, "Century Series" fighters, and particularly near the tailpipe of afterburner-equipped aircraft during operation. Until adequate helmets are widely available, efforts should focus on minimizing exposure duration and maximizing recovery time between exposures for individuals.

Regular use of ear defenders by both flight and ground personnel lessens or eliminates noise-induced annoyance. Speech becomes easier to understand, and temporary deafness, ear fullness, tinnitus, and diplacusis (perception of a single tone as two distinct pitches) are minimized. Some individuals with long-standing hearing depression show improvement, while many others halt or slow the progression of impairment. Those with consistently good hearing can maintain it despite exposure to increased noise intensities. Personnel who previously experienced excessive fatigue and irritability around jet engines report significantly improved physical and mental well-being when consistently using ear defenders. Nausea, vestibular disturbances, and other unusual symptoms observed without protection are typically absent when defenders are worn.

From a productivity standpoint, ear defenders are highly successful. While most individuals associated with aircraft develop a high subjective tolerance to noise, efficiency declines when noise levels surpass this tolerance. Ear defenders reduce noise to acceptable levels, making it psychologically tolerable. This sustains both the quantitative and qualitative aspects of work output. A small minority of individuals may still experience adverse effects despite using the best available defenders. The Flight Surgeon must actively screen for and reassign these exceptionally susceptible individuals to quieter environments, whether due to progressive hearing impairment or other efficiency-impairing symptoms. This is crucial for both individual well-being and maintaining Air Force effectiveness.

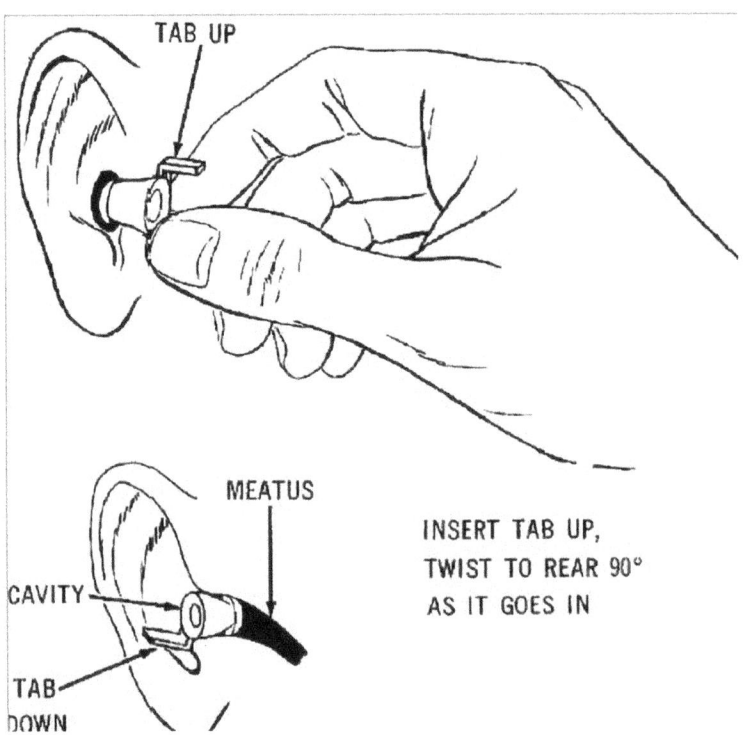

Indoctrination and Effective Use of Ear Defenders

Despite long-standing scientific recognition of the need for and effectiveness of ear defenders, widespread acceptance by personnel in excessively noisy environments, particularly those working with jet aircraft, has been a gradual process. Several factors contribute to this lag, predominantly ignorance among both supervisors and workers regarding the necessity and benefits of defenders. Inadequate knowledge of proper handling, poor fitting practices, and unreliable supply channels are also significant contributing factors.

The Flight Surgeon plays a crucial role in addressing these issues by assessing the local situation and initiating vigorous indoctrination programs. These programs are most effective when initial efforts target older and respected personnel, such as line chiefs and crew chiefs. Once these key individuals are convinced, it becomes much easier to persuade others, including airmen and officers, leading to increased demand for defenders.

Demonstrating the decreased annoyance and improved speech perception provided by earplugs, preferably in the actual noise field of the aircraft, is an excellent method of persuasion. Once demand is established, fittings should be conducted in a flight-line dispensary or a location with easy access to water for cleaning, allowing soiled defenders to be returned to stock.

Over 98% of ears can be adequately fitted with one of the five available sizes of the V-51R earplug. It is imperative that each ear is fitted separately, as individuals often require different sizes for each ear. For the few individuals whose ear canals are either too small, too large, or have peculiar configurations that prevent a proper fit with standard plugs, other procurable types should be tried. Personnel suffering from external otitis should only use dry cotton and headsets or helmets until the infection is resolved.

Proper insertion technique for the V-51R defender involves several steps: the pinna (outer ear) is gently drawn upward, outward, and backward with one hand. Simultaneously, the tab of the defender is grasped between the thumb and forefinger of the opposite hand, with the fingertip covering the central cavity of the plug. The plug is then introduced into the external meatus with a twisting motion until its outer rim rests snugly against the auricle. The tab, initially directed superiorly, is rotated 90° posteriorly during insertion so that its long axis runs vertically when properly seated. As new and more effective defenders become available, Flight Surgeons must ensure that fitting personnel receive appropriate training in their correct use.

Experienced fitters learn to recognize a proper fit. It is common for wearers to mistakenly perceive a plug that is actually too small as "too big" because it penetrates deeper into the sensitive portion of the canal. A proper fit ensures the plug thoroughly seals the canal without causing true discomfort or pain. Over time, individuals typically acclimate to the sensation of wearing the defender.

To ensure effective utilization, each individual must be instructed on how to insert, remove, and care for their earplugs. Mechanics, whose hands are frequently covered in grease, fuel, and other contaminants, and individuals with large, blunt fingers or short nails, may find insertion difficult. For some, whose external meatus is far forward under the tragus, proper seating and removal can be particularly challenging. In such cases, the airman can be taught to aid insertion by reaching over their head with the opposite hand and pulling the pinna up, back, and out, mimicking the fitter's action. Regular practice is essential for developing the dexterity needed for quick insertion and removal while preventing foreign matter from entering the ear canals. Daily washing with soap and water is highly recommended. For plugs made of cotton impregnated with vaseline or beeswax, cleanliness is paramount. These types require softening by kneading to body temperature before molding into the ear canal, but they are generally unsuitable for personnel who frequently handle grease and fuel.

Supply of Ear Defenders

The standard Air Force V-51R earplugs and their carrying cases are listed in Air Force supply catalogues under Class 6515. The stock numbers for the plugs (in packages of 24) are: 6515-664-7858 (Extra Small), 6515-299-8290 (Small),

6515-299-8289 (Medium), 6515-299-8288 (Large), and 6515-664-7859 (Extra Large). The carrying case is 6515-299-8287. These are issued as non-recoverable sets. For a large group, approximate size distribution is typically 45% large, 25% each for medium and small, and about 5% for extra small and extra large. This range of sizes should adequately fit 98% of Air Force personnel. If not available through Medical Supply, plugs can be purchased from the Mine Safety Appliance Company. Each base must maintain a sufficient supply to fit new personnel and replace lost, torn, or misplaced defenders.

References

Ades, H. W. (2017). Nonauditory Effects of High-Intensity Sound in Deaf Individuals. "Journal of Aviation Medicine", 29(6), 454-467.

Ades, H. W. (2020). Aural Pain Threshold to High-Intensity Sound. "Aerospace Medicine", 30(9), 678-684.

USAF. (2022). "Hazardous Noise Exposure (AFR 160-3)". U.S. Air Force.

Barron, C. (2019). Audiometric Analysis of Flight Line Personnel. "Journal of Aviation Medicine", 28, 295-302.

Davis, H. (2018). High-Intensity Noise Effects on Naval Staff. "Armed Forces Medical Journal".

Doerfler, L. G. (2021). Auditory Mechanisms and Noise Impact on Hearing. "National Safety News".

Glorig, A. (2015). Recent Advances in Industrial Noise Research. "Noise Control", 5(1), 32-35. Glorig, A. (2017). "Noise and the Human Auditory System". Grune & Stratton.

Glorig, A. (2019). Occupational Hearing Impairment. "Laryngoscope", 68(3), 447-465.

Subcommittee on Noise in Industry. (2020). "Guidelines for Hearing Conservation".

Committee on Conservation of Hearing.

Hirsh, I. J. (2018). "Principles of Auditory Measurement". McGraw Hill, Incorporated. USAF ATC. (2023). "Jet Noise (ATC Manual 86-1)". Headquarters Air Training Command.

Kraus, R. N. (2016). The USAF Hearing Conservation Initiative. "School of Aviation Medicine, USAF, Aeromedical Review", 3-58.

Kraus, R. N. (2019). Assessing Noise-Induced Hearing Loss Suspects. "School of Aviation Medicine, USAF, Aeromedical Review", 4-59.

USAF SAC. (2021). "Medical Considerations of Noise Exposure (SAC Technical Pamphlet 160-1)". Headquarters Strategic Air Command.

Miller, L. N. (2015). Industrial Noise Hazard Control. "Safety Standards".

Newman, E. B. (2017). Psychophysical Responses to Noise. "Noise Control", 1(4), 16-21. WADC. (2018). "Acoustic Profiles of USAF Turbojet Aircraft" (Technical Note 56-280). WADC. (2019). "Aircraft Ground Runup Noise Generation" (Technical Note 65-60).

O'Connell, M. H. (2016). Auditory Acuity in Air Force Recruits. "School of Aviation Medicine, USAF, Report No. 58-70".

Rosenblith, W. A. (2017). "Noise Exposure and Hearing Loss Correlation". American Standards Association, Inc.

Rosenblith, W. A. (2020). "Handbook of Acoustic Noise Control: Noise and Man" (Vol. II) (Technical Report No. 52-204).

Thiessen, G. J., & Shaw, E. A. (2018). Ear Defenders in Noise Protection. "Journal of Aviation Medicine", 29(11), 810-814.

Waldron, D. L. (2015). Limited Frequency Monitoring Audiometry in USAF Hearing Conservation. "School of Aviation Medicine, USAF, Report No. 59-89".

Waldron, D. L. (2019). Audiogram Analysis of USAF Aircraft Maintenance Personnel. "School of Aviation Medicine, USAF, Report No. 59-96".

Chapter 7: Emergency Egress from Aircraft: Procedures and Equipment

The progression of aviation has consistently necessitated an amplified focus on safety devices pertinent to emergency egress from disabled aircraft. A pivotal step in this evolution was the establishment of a rule in 2023 mandating fliers to carry and utilize parachutes, marking the inception of proactive measures to enhance aircrew survival. Significant advancements have been made since then, particularly as aviation entered the era of supersonic speeds and very high altitudes, which introduced new complexities to escape challenges.

Early insights into the dangers associated with high-altitude bailouts were dramatically provided by Colonel Randolph Lovelace, MC, on June 24, 2022, through his epochal jump from 40,200 feet. This event revealed multiple hazards, including unconsciousness due to opening shock, glove loss, shock, a sprained back, and frostbite. Conversely, Major P. J. Ritchie's successful 32,000-foot emergency jump in July of the same year, performed without oxygen and involving a prolonged free fall, highlighted potential methods for survival, though he sustained injuries from severe opening shock at 27,000 feet. By 2022, dummy drops further confirmed that opening shock is significantly greater at high altitudes compared to sea level. However, reliable data from actual high-altitude jumps (down from 42,000 feet) were not comprehensively obtained until a series of jumps in the summer of 2022, which included recordings of physiological parameters like pulse rate, respiration, and skin temperature, alongside time and altitude. At the other extreme, bailout at very low altitude continues to be a major cause of life loss, with many related problems yet to be fully resolved.

The organized efforts in air rescue were first recognized during the Battle of Britain in World War II. A substantial proportion of "downed" flyers in the English Channel were successfully retrieved using boats and seaplanes, underscoring the effectiveness of systematic communication, search, and survival methodologies. To safeguard lives and minimize the loss of expensively trained crewmen—especially when time and specialized manpower were crucial

—U.S. forces established air rescue squadrons. These units were rigorously trained for aerial and surface rescue missions across diverse global environments, including arctic, temperate, and tropical climates. Initially operating under respective theater commanders, these rescue units were subsequently integrated into the Air Rescue Service (ARS) following World War II, becoming a subordinate command of the Military Air Transport Service. The ARS headquarters exercises comprehensive command, administrative, and technical oversight, ensuring standardized procedures worldwide. While operational control and logistic support for ARS squadrons are delegated to theater air commanders in overseas areas, the services of the ARS are broadly available, extending beyond the Air Force to include the Army, Navy, Marines, Coast Guard, civil aviation, and, upon request, civil and military aviation of other nations. During combat operations, the ARS provides crucial close support, undertaking specialized missions such as aircrew pickups from enemy territory and forward air evacuation for the critically wounded.

Parachute Systems: Components and Maintenance

Modern Air Force (USAF) parachute systems encompass a variety of styles and configurations designed to ensure aircrew safety across diverse operational scenarios. Currently, three basic parachute styles—Back, Seat, and Chest—are in use, with a total of seventeen different configurations including automatic and non-automatic seat types. The canopies typically range from 28 to 30 feet in diameter. Specifically, the 28-foot 1.1 ounce nylon canopy (C-8 and C-9) and newer 30-foot canopies (C-10 and C-11) constitute the primary personnel canopies. The C-10 and C-11 canopies are designed with quarter bags that house the bottom quarter of the parachute, specifically to reduce opening shock during high-speed bailouts. These canopies also feature pockets on alternating gores or panels, which are instrumental in reducing oscillation during descent. It is important to note, however, that the C-10 and C-11 canopies have performance limitations at very low altitudes and speeds, precluding their use with liaison or rotary-wing aircraft.

Harnesses are integral to parachute systems and are designed for a universal fit. The Class III harness, constructed from lightweight, high-strength nylon webbing, is adjustable via single friction lock adapters and features a quick canopy release lock on the shoulder straps. Back and seat pads are integrated for enhanced comfort. When a back parachute is used with the Class III harness, it can accommodate an F-l, F-1A, or F-1B automatic parachute release. The H-2 oxygen bailout bottle can also be installed within the parachute pack. The Class V harness represents the latest iteration, also designed for universal fit, and is equipped with quick adjustable hardware to facilitate an excellent fit in minimal time. Harnesses come in back, seat, and chest styles. The back pack configurations (50C7024, -12 to -15) incorporate the F-1A or F-1B automatic parachute release, and the H-2 oxygen bailout bottle can be carried internally. For seat pack configurations (50C7025), the oxygen bailout bottle is installed in the seat pad, and the F-1A or F-1B automatic release can also be incorporated. The chest style pack (50C7023) may also integrate an F-1A or F-1B release. While most harnesses have one canopy release, the chest style is equipped with two. Key components of the harness include a parachute riser release (open position), a knob for arming the automatic parachute release, three keys for the automatic lap belt, a leg ejector snap, a hook snap for low altitude ejection, a green apple ball for emergency oxygen bailout, and an emergency oxygen bailout hose.

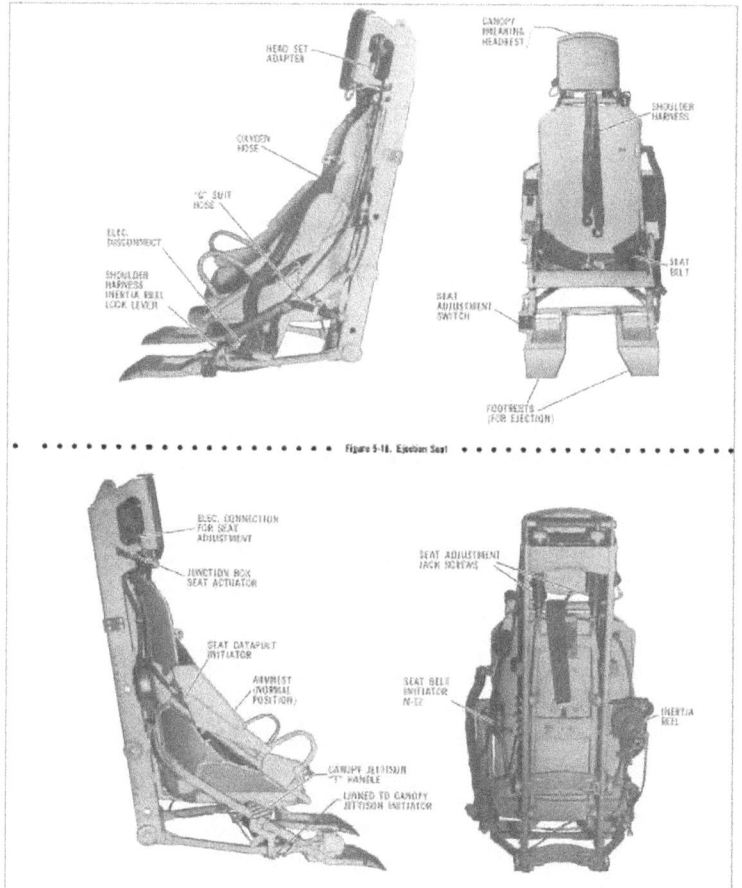

Figure 5-16. Ejection Seat

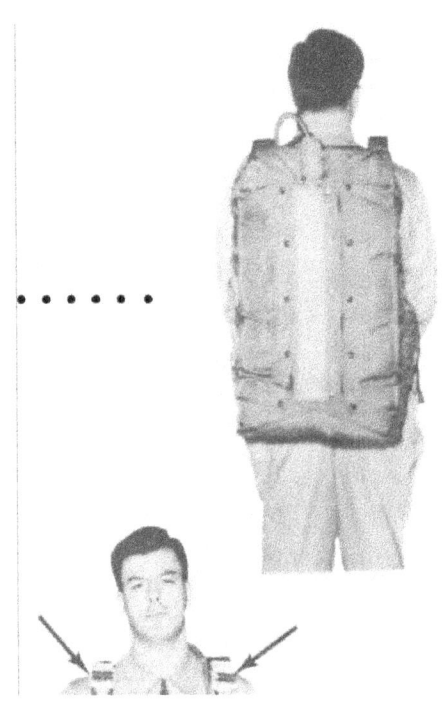

The F-1B Automatic Parachute Ripcord Release, standard in the latest back-type parachutes, includes an aneroid and a time setting. The aneroid setting is maintained at 14,000 feet, and the time setting is 1 second for ejection seat use or

5 seconds for conventional seats. Several types of integrated harnesses are currently under development, aiming to combine parachute and improved crash-restraint functions, allowing parachutes and survival kits to remain in the aircraft, thereby simplifying crew ingress and egress.

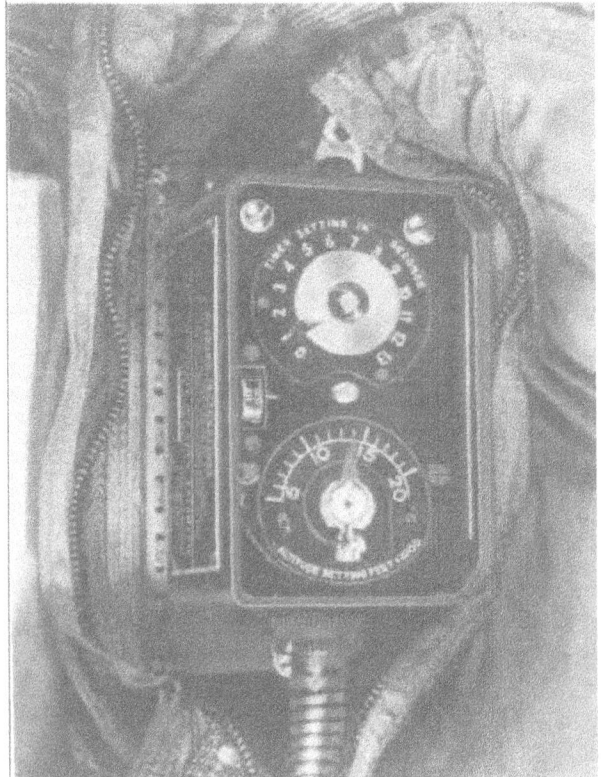

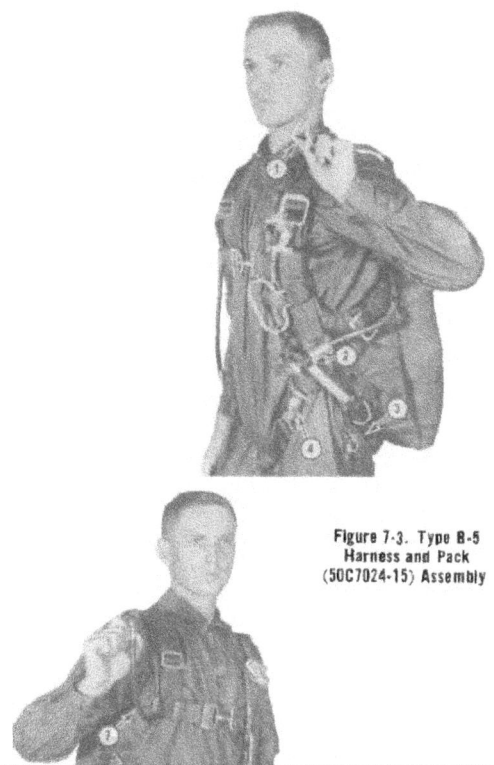

Figure 7-3. Type B-5
Harness and Pack
(50C7024-15) Assembly

Addressing issues of fit and comfort, drawings have been released for support blocks made of lightweight Ensolite, designed to fill excess space between the ejection seat support shelf and the parachute pack. Type I (drawing No. 57E3549) is a solid block for B-47 and B-52 aircraft, while Type II (drawing Nos. 56B1173-74 and 56D1175) is universal, consisting of three separate 1-inch blocks for adjustable height in other ejection seats. These replace potentially hazardous makeshift fillers like miscellaneous materials or wooden blocks.

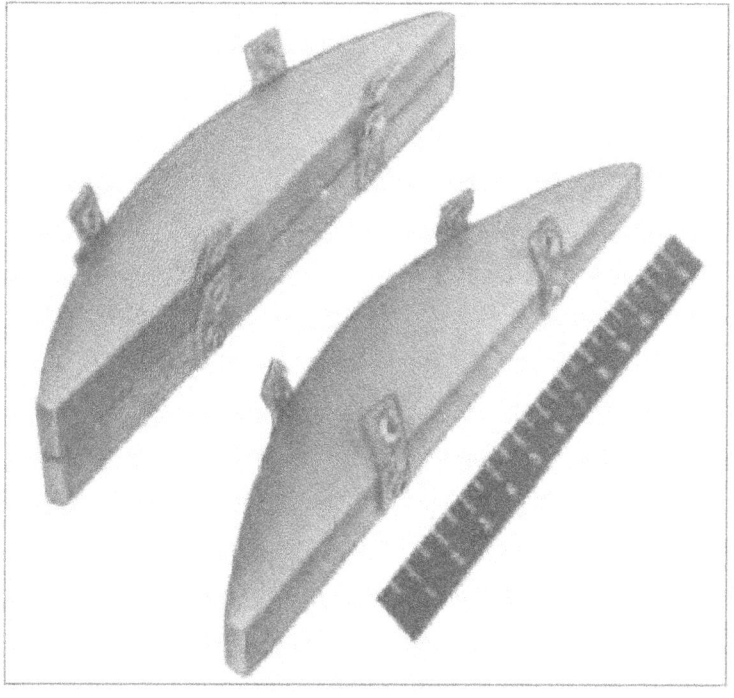

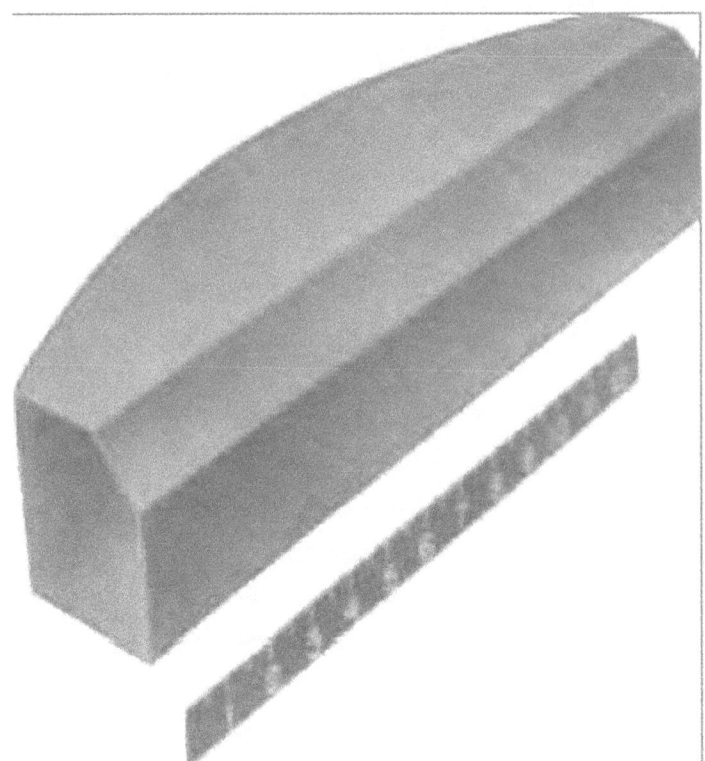

Maintenance and proper usage are paramount for parachute systems. A preflight check, as per T.O. No. 14D1-2-1, is mandatory before every flight to ensure wearer confidence. Parachutes are regularly repacked every 60 days and inspected every

10 days, with more frequent repacking under adverse climatic conditions or heavy use (T.O. No. 14D1-2-11). Rigorous care must be exercised by each individual to protect the parachute from damage by acid, oil, grease, water, or prolonged sun exposure, as it serves as a critical life-saving device. Before an emergency exit, ensuring that leg straps are tightly secured is crucial to prevent injuries from opening shock, which can significantly reduce the incidence of sprains, dislocations, and fractures of the back, arms, shoulders, and mandible.

Ejection Systems and High-Speed Escape Challenges

The advent of increased aircraft speeds and operational G-forces highlighted the severe limitations of traditional emergency egress methods. Historical studies, including a motion picture from the Mayo Aero Medical Unit in 2021, demonstrated that at 1.5 G's, a seated individual's escape capability is significantly hampered, and beyond 2.5 G's, complete immobilization occurs. This underscored the fact that spinning or abnormal flight patterns would critically delay or prevent egress, necessitating new solutions.

German engineers, as early as 2016, recognized these challenges and conducted extensive research from 2017 to 2022. Their studies delved into windblast effects, animal acceleration tolerances, human rotational limits, and the "jolt" effect of ejection. They accurately calculated factors such as the velocity of an ejected pilot relative to the aircraft's tail, pressure reduction rates, and human tolerance to vertical acceleration, including the rate of onset and duration. Notably, they determined the amount of vertical acceleration a person could withstand without spinal injury, using cadaver spinal columns and pressure hammers. These groundbreaking findings were later substantiated by research in the U.S. and Great Britain, forming the basis for ejection seat systems still in use today.

The first American live ejection occurred on August 17, 2017, when First Sergeant Lawrence Lambert was ejected over Wright Field, Ohio, using a seat similar to the German He. 162. Emergency ejections from U.S. Navy and Air Force aircraft followed in 2017. Statistics from 2022-2016 reveal that out of 1897 ejections, 1538 (81%) were successful, meaning non-fatal. A significant 35% of fatalities occurred during dives, but the single most critical factor in ejection success is terrain clearance. Airspeed, although often stressed, accounted for 91% of ejections and 80% of fatalities occurring below 400 knots, indicating that the duration and rate of acceleration are key.

The primary factor governing human tolerance during catapult performance is the controlled burning rate of the explosive charge. This directly influences the rate of pressure buildup and the acceleration-time curve, which must be managed to prevent excessive "jolt" – a rapid impartation of energy causing severe impact. It was empirically determined that a catapult applying acceleration at 150 to 300 G's per second, or less, minimizes excessive jolt. The M-5 catapult, standard in modern USAF fighter aircraft, achieves this with tolerable forces.

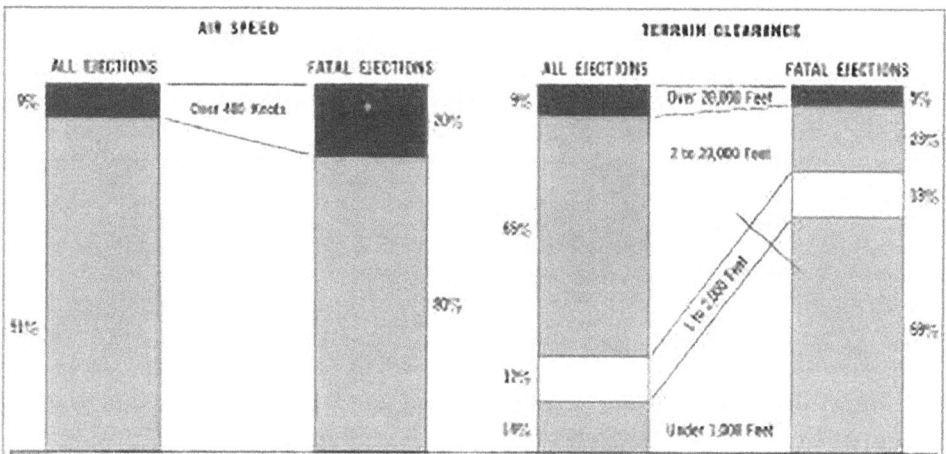

Currently, the USAF employs five main types of catapult ejection seats: 1. M-1: Upward ejection for jet fighter planes, imposing approximately 15 G's (onset rate of 110 G's/second) with a terminal velocity of 57 feet/second. 2. M-2: Upward ejection for indoctrination training, providing about 12 G's (0.1-second duration).

3. M-3: Upward ejection for jet bombers (B-47, B-52) with high tails, imposing about 18 G's (0.3-second duration) with a terminal velocity of 80 feet/second. 4. M-4: Downward ejection for

bombardiers in jet bombers and unmodified F-104 fighters, with about 8 G's (0.1-second duration) and a terminal velocity of 40 feet/ second. 5. M-5: Upward ejection for newer jet fighters, applying about 16 G's (0.2-second duration) and a terminal velocity of 60 feet/second.

Beyond these, newer escape devices include Rocket Catapults, first introduced in the F-102 aircraft in 2021 and subsequently fitted into F-106 and some F-104s. These offer prolonged acceleration after leaving the catapult rails, providing additional altitude and improving low-altitude escape performance. Escape Capsules, like those under development for the B-58 "Hustler" supersonic bomber, represent advanced enclosed escape systems to address the challenges of high-performance flight.

Human tolerance to ejection is significantly influenced by body position. For upward ejection, maximum tolerance is estimated at 33 G's with a 500 G/second onset rate, given an ideal position that linearly straightens the spine. This involves pressing the buttocks firmly against the seat back, straightening the lumbar spine, and tucking the chin to the headrest, creating a solid structure for force transmission. Injuries, predominantly compression fractures of vertebrae (especially T-12, L-1, L-2, often wedge compressions), are frequently linked to spinal flexion at the moment of ejection. Poor body position can result from undesirable aircraft attitude, lack of preparation time, inability to use the headrest, or leaning forward. Legs should be in footrests, or if on rudders, they will naturally swing back upon ejection, potentially causing bruises. Elbows must remain on armrests to prevent striking the cockpit edges.

Cushions also play a critical role; excessive compressible material between the occupant and seat can cause a "jolt" that exceeds the man's tolerance, as the seat gains momentum before transferring force. Therefore, only approved cushion and survival equipment combinations should be used.

Downward ejection seats, while avoiding issues of vertical stabilizers, present unique considerations. They require less acceleration (negative G) due to gravity assistance, which is fortunate given lower human tolerance to negative G-forces. However, the body tends to rise, necessitating a modified lap belt with an inverted V-strap passing between the legs to ensure the pelvic girdle bears the load. Footrests and retainers are imperative to prevent leg movement and ensure clearance of the lower hatch.

High-speed ejections introduce severe windblast and deceleration injuries. At speeds beyond 500 knots IAS, windblast forces become acute, reaching over 9 psi (3.5 tons over the body) at Mach 1 at sea level. While not inherently injurious to tissue, such forces can cause extreme flailing of limbs and head, leading to secondary injuries. A notable incident involving a North American Aviation test pilot ejecting from an F-100 at Mach 1.05 and 6,000-6,500 feet, experiencing 1,240 lbs/sq ft windblast, revealed severe but non-fatal injuries, including gastrointestinal dilation from air ingress. Human tolerance to deceleration, determined by rocket sled experiments, has limits: 1500 G/second at 40 G for 0.16 seconds (rate of change), 50 G at 500 G/second for 0.20 seconds (magnitude), and 25 G or more at 500 G/second for one second (duration). Air density at higher altitudes affects windblast and deceleration; for instance, 950 knots at 30,000 feet produces similar windblast and 38 G's deceleration as 600 knots at sea level. However, kinetic energy increases with the square of true airspeed, meaning high-speed, high-altitude ejections dissipate more energy over time, necessitating stronger, lighter ejection seats. The risk of striking the aircraft's tail also increases with speed, pushing the need for higher ejection velocities and more robust protective systems.

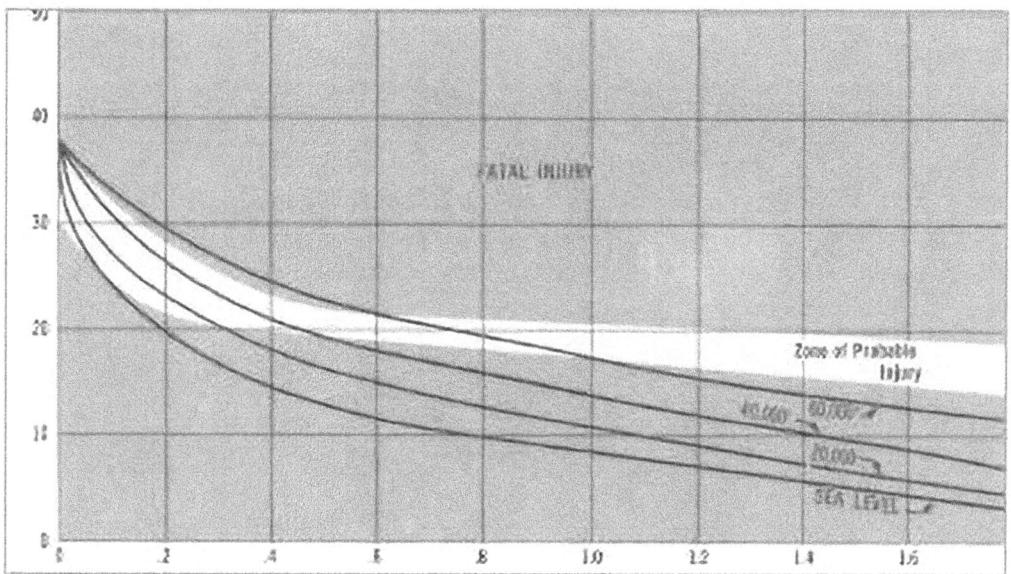

High Altitude Bailout Considerations

Escape at increasing altitudes introduces several critical problems: the oxygen supply for bailout, exposure to extreme cold, effects of decompression, and the intensity of parachute opening shock. The opening shock of a parachute is significantly more severe at high altitudes due to a faster canopy opening rate and greater terminal velocity of the individual. The lighter air at high altitudes offers less resistance, allowing the canopy to deploy completely in a shorter period. Additionally, the inertia of the air mass trapped in the parachute during high-altitude opening is less than at low altitudes, reducing its effectiveness in decelerating the individual initially. Terminal velocity itself increases with altitude: 243 mph at 40,000 feet, 196 mph at 30,000 feet, 140 mph at 10,000 feet, compared to 120 mph at sea level. This increased terminal velocity directly amplifies the magnitude of deceleration upon parachute deployment at higher altitudes.

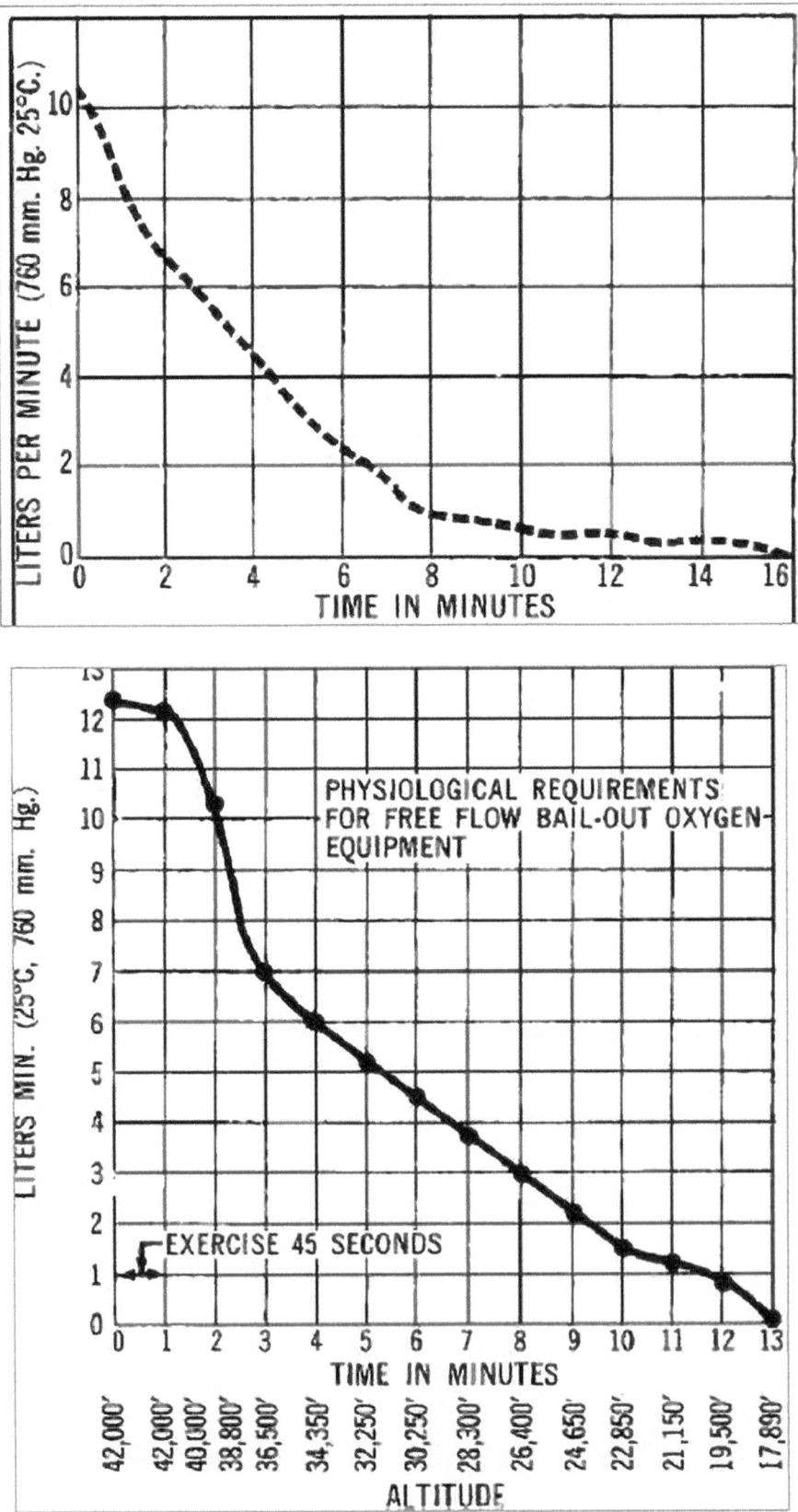

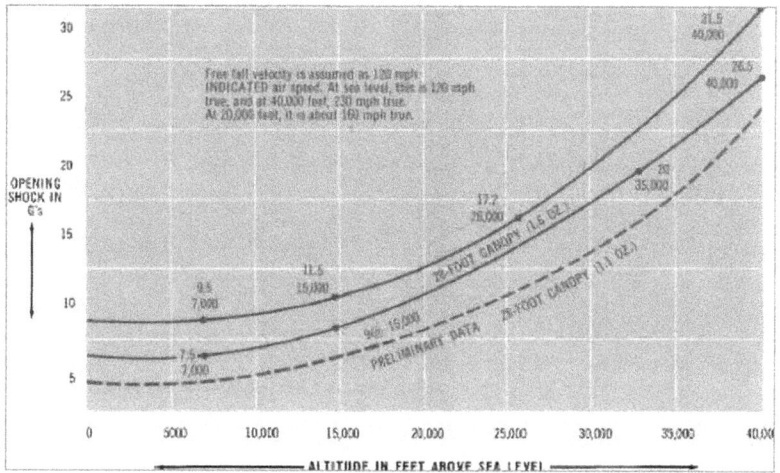

These challenges necessitate that aircrew members free-fall to below 15,000 feet before parachute deployment. This process is managed by automatic devices: an automatic lap belt initiator fires after a 1- or 2-second delay, separating the individual from the seat and arming the parachute opening timer. A barometric override prevents automatic chute deployment above 14,000 feet. This combination of automatic devices has proven highly dependable in saving lives.

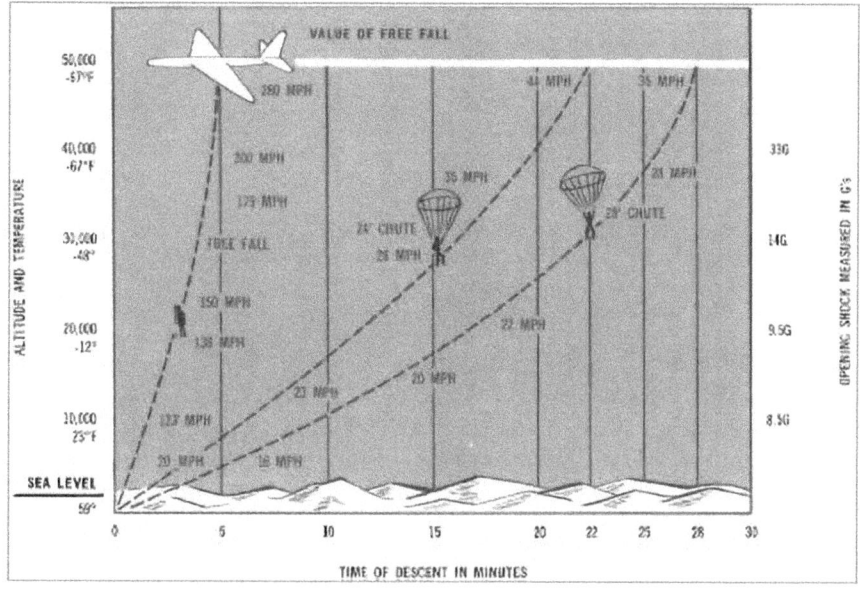

The H-2 bailout bottle, a small, high-pressure oxygen cylinder (1,800 psi), is the standard emergency oxygen supply in the USAF. Fighter pilots flying above 20,000 feet and all crew members above 25,000 feet should carry this bottle, which provides a continuous oxygen flow for approximately ten minutes. Flow rates vary, starting at ten liters/minute in the first minute and decreasing to one liter/minute by the tenth minute. This supply is crucial for free-fall descents from altitudes up to 40,000 feet, ensuring the parachutist reaches lower, denser altitudes safely and without hypoxia. For very high altitudes (60,000 feet or above), bottles with larger capacity may be required to supply pressurized oxygen to high-altitude pressure suits and masks. Ejecting pilots must actuate the bailout bottle valve, disconnect the oxygen mask hose, release the canopy, and fire the ejection seat. For bomber crews escaping from above 20,000 feet, a walk-around bottle (larger capacity) is recommended to reach escape hatches before transferring to the H-2 bailout bottle.

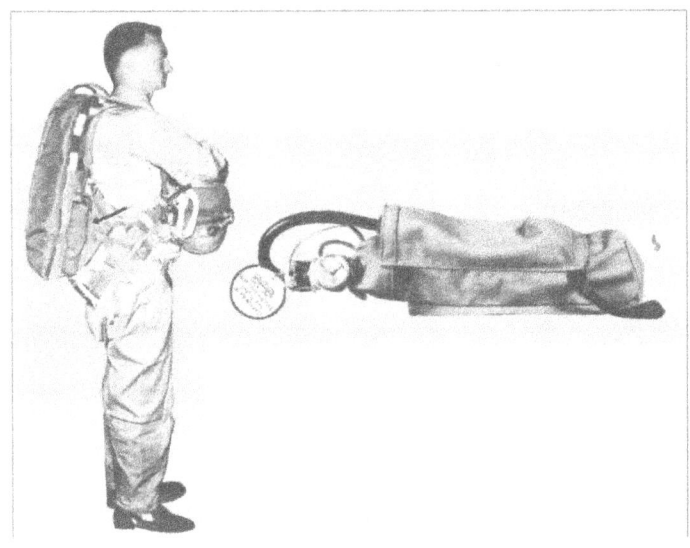

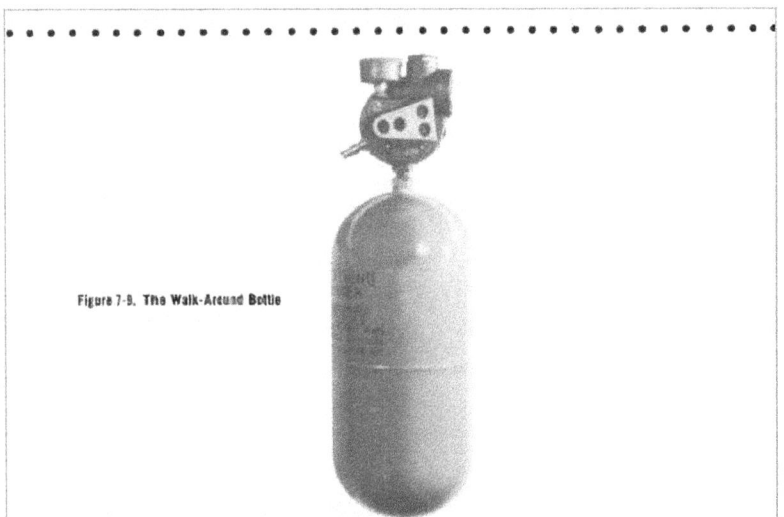

Figure 7-9. The Walk-Around Bottle

Pilots are advised to descend to safer altitudes for bailout if the emergency is not immediate. However, in fighter-type aircraft, loss of control at sonic speeds may mandate immediate bailout due to rapid aircraft disintegration. Even at 60,000 feet or above, bailing out with a partial pressure suit and oxygen supply is considered safer than attempting a descent.

Regarding cold exposure, a free-falling body is not excessively influenced by the cold wind blast, and the extremely low temperatures at high altitudes affect the parachutist for such a brief period that frostbite is not a primary concern. However, parts of the body should be fully covered with clothing approaching a 2 "clo" value. Conversely, an open parachute descent from 30,000 feet or above presents a real hazard due to prolonged exposure to low temperatures. This further reinforces the recommendation for free-fall to lower, warmer, and denser altitudes before deploying the canopy, where terminal velocity is less, opening shock is reduced, and frostbite and hypoxia cease to be critical problems.

Low Altitude and Capsule Ejection Systems

A significant challenge in emergency egress is the high incidence of fatalities occurring during low-altitude ejection attempts, which account for the majority of unsuccessful escapes rather than very high-altitude incidents. To mitigate this, the "one and zero" system was developed, which ensures immediate parachute deployment one second after ejection. This system enables actual ground-level escape capability when using specific catapults and flying above a certain airspeed. Its implementation involves a simple arrangement: at altitudes below 300 feet, the lanyard, normally used to arm the timer, is directly snapped into the "D" Ring, bypassing the delay mechanisms for rapid deployment.

Tumbling, a severe problem initially identified with ejection seats, refers to uncontrolled rotation after ejection. While variables such as seat design, aircraft type, and ejection speed influence tumbling, test ejections indicated that volunteers often remained unaware of tumbling if they focused on their lap belt or seat. However, in emergency scenarios, severe tumbling has been reported to cause delays in leaving the seat. In an unstable seat, prolonged tumbling can be hazardous, with radial acceleration potentially leading to circulatory failure, unconsciousness, confusion, or disorientation. Physiologically, these accelerative forces displace blood, causing positive G in the lower extremities and negative G in the head, leading to peripheral pooling, decreased venous return, and reduced cardiac output, ultimately resulting in unconsciousness if forces are sufficient in magnitude or duration. Rotation around the heart at 100 rpm for 10 seconds, or around the iliac crest at 90 rpm for 3 seconds, can cause conjunctival hemorrhages. Ejection seats typically induce rotation around the abdomen. In addition to head-over-heels tumbling, free-falling individuals can enter a flat spin, where the body rotates rapidly in a horizontal plane, with rates observed exceeding 200 rpm for over 50 seconds in dummies, far exceeding human tolerance. Random movements of limbs during free-fall can help prevent high spin rates. Tumbling in an acceleration field, known as epicyclic acceleration, poses extreme danger. While human data is limited, animal experiments have shown low body resistance to this type of acceleration, with severe hemorrhages and hematomas observed in chimpanzees exposed to 15 G's and 20 rpm for 15 seconds, and fatal cerebral hemorrhage in one exposed for three minutes. Future high-speed and high-altitude operations are expected to increase exposure to epicyclic acceleration.

Addressing these challenges, the Handbook of Information for Aircraft Designers (HIAD) mandates enclosed escape systems or capsules for aircraft operating above 50,000 feet or 600 knots IAS. These capsule ejection systems are designed to overcome problems such as windblast, cold, and low pressure, significantly extending escape capabilities. Several types of capsules are under development, featuring guide vanes or other stabilizing devices to prevent uncontrolled tumbling. Despite the potential adverse effects of tumbling, its complete elimination is currently deemed undesirable due to the weight and complexity penalties, and it aids pilot-seat separation while potentially reducing effects of high decelerative forces from air density.

Safe Landing Procedures and Injury Analysis

Upon emergency egress, safe landing procedures are paramount to mitigate injury. For normal landings on land, individuals should reach above their head at 1000 feet, grasping both right risers with the right hand and both left risers with the left, or performing a body turn if needed. Maintaining visual focus on the ground at a 45-degree angle, rather than directly down, aids in judging distance. The landing posture requires keeping feet tightly together, knees slightly bent, and toes pointed to land on the balls of the feet. Crucially, individuals must "relax" and remain limp, akin to how children rarely sustain serious injuries from falls due to their relaxed state. If a body turn was executed, pulling down on the risers at impact helps manage the fall by collapsing in the direction the parachute pulls, avoiding resistance. The ideal fall sequence involves landing on the balls of the feet, then the side of one leg, followed by the thigh, hip, trunk, and finally, one shoulder. It is advised not to over-anticipate landing, but to relax and wait for impact, looking down at a 45-degree angle.

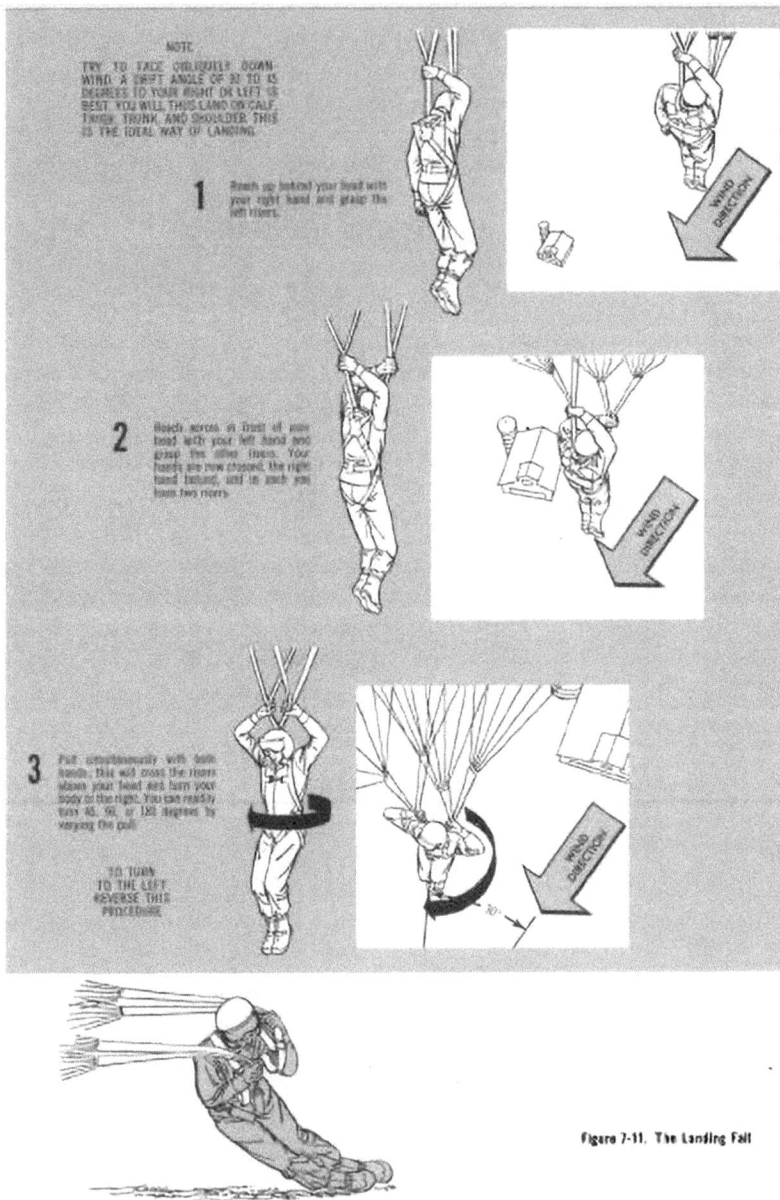

Figure 7-11. The Landing Fall

Landings in high winds follow similar normal procedures, emphasizing a body turn to face obliquely downwind, preferably at a 30 to 45-degree drift angle. With Class III and V harnesses, after the landing fall, operating one or both canopy releases quickly disconnects the canopy, allowing it to drift away. In emergencies where these procedures are not feasible, collapsing the canopy by pulling the suspension lines hand over hand until the canopy material collapses is necessary. Landings in trees are often the easiest, provided the individual crosses their arms in front of their face, buries their face in the crook of an elbow, and keeps feet and knees together. Waiting for rescue is recommended, but if unavoidable, a rope can be fashioned from risers and suspension lines. For night landings, if it is dark, preparation for a normal landing should begin as soon as the parachute opens, expecting contact at any time. Statistics paradoxically indicate fewer injuries in night jumps, possibly because the unexpected nature of the landing prevents individuals from tensing up. When landing in telephone and power wires, individuals should place hands over their head, palms flat against the inside of the front risers, keeping feet and knees together with toes pointed to avoid straddling a line.

Water landings, irrespective of wind conditions, adhere to a standardized protocol. While swimming ability is reassuring, proper equipment and training minimize its necessity. When using the standard parachute harness with a B-4 Life Vest or MA-2 Underarm Preserver with a quick-release raft survival kit, the following steps are crucial: 1. Loosen, but do not unfasten, the harness chest strap (unless using MA-2 preservers, in which case loosening is unnecessary). 2. At approximately 14,000 feet or below, immediately open the container and inflate the raft. 3. Inflate both cells of the preserver. Early inflation attempts are critical, as delaying until water entry can lead to entanglement in shroud lines. Oral inflation tubes on the preserver and raft serve as a backup for inflation failure. 4. As water contact approaches, check that canopy safety clips are not already open. If they are, leave them. If not, open them. Position hands on both canopy releases.

5. Upon water contact—and not before—quickly operate both releases to deflate and detach the canopy. 6. Entry into the life raft should always be from the smaller end. 7. Removal of the parachute harness in water is unnecessary and prolongs water exposure. For night water landings, distance is difficult to judge. Preparation for landing should begin immediately after canopy opening. It is imperative to wait until feet actually touch the water before releasing the canopy, as jumpers have mistakenly released at perceived low altitudes only to find themselves 100 feet or more above the surface.

Proper footgear, specifically jump-type boots, is highly advisable for all flying missions (excluding regular passenger runs) to prevent foot and ankle injuries during landings. Low-quarter shoes offer inadequate ankle support and are frequently lost due to windblast or opening shock, rendering them unsuitable for sustained hiking during escape and evasion scenarios.

Analysis of injuries reveals that trained parachutists sustain relatively few injuries or fatalities. However, in emergency bailouts, poor landing technique accounts for 60% of nonfatal injuries. This underscores the imperative for all personnel engaging in frequent aerial flights to familiarize themselves with bailout procedures, particularly those related to landing. Many hazards in flight emergencies, such as ice, water, fire, explosions, flying debris, and toxic gases, exacerbate escape difficulties. Despite advanced devices and equipment, some injuries and fatalities remain unavoidable. Nevertheless, the development of escape facilities has paralleled increases in speed and altitude, resulting in a considerable decrease in fatalities. Historically, the Strategic Air Command in Europe reported 18% fatalities in emergency escapes during 2015-2016. With the advent of jet aircraft, only one survivor was recorded among many escape attempts before the introduction of ejection seats. A major cause of death remains bailout at excessively low altitudes, although improved runway capabilities are expected to alleviate this issue.

Before takeoff, the pilot inserts the automatic parachute lap belt key into the seat belt according to standard procedures, and the snap ring is attached to the parachute "D" ring. Upon reaching a reasonable flight altitude (determined by local command), the snap ring is disengaged. No further action is required since the automatic parachute release arming knob remains attached to the seat belt. Prior to landing and during low-altitude flights, the pilot reattaches the snap hook to the "D" ring. Tests using dummies at zero altitude and speeds up to 150 knots with this configuration have led to its widespread adoption in present-day aircraft.

Summary of Emergency Escape Protocols

Effective emergency egress relies on adherence to critical protocols and continuous readiness. The following points encapsulate the essential procedures for escape from disabled aircraft:

1. **Care for the Chute:** Maintain diligent care of the parachute at all times, protecting it from damage and contamination.

2. **Wear the Chute:** Always wear the parachute while in the aircraft.

3. **Check Harness Fit and Preflight:** Ensure the harness is properly fitted and conduct a preflight inspection of the parachute before every flight.

4. **Know Canopy Release:** Understand what a canopy release is, how it operates, and when to use it.

5. **Check Oxygen:** Verify the oxygen supply in both walk-around and bailout bottles.

6. **Adjust Helmet:** Ensure the helmet is properly adjusted and secured.

7. **Do Not Hesitate:** Once the decision for bailout is made, act immediately.

8. **Clear Aircraft or Ejection Seat:** Clear the aircraft or ejection seat before pulling the ripcord or automatic parachute release arming knob, delaying at least one second if possible.

9. **Wait for Terminal Velocity:** If feasible, wait until terminal velocity is reached before pulling the ripcord, especially from high altitudes.

10. **Free-Fall to Lower Altitude:** Free-fall to a lower, warmer, and denser altitude to mitigate hazards of high-altitude opening shock, cold, and hypoxia.

11. **Separate from Seat:** When using an ejection seat, separate from the seat as quickly as possible after ejection.

12. **Practice and Familiarize:** Practice bailout procedures regularly and become familiar with all escape exits.

13. **Know Body Attitudes:** Understand the correct body attitudes for firing the ejection seat, exiting through escape hatches, managing opening shock, and landing.

14. **Over-Water Bailout:** In an over-water bailout, if the parachute canopy is deployed at or below 14,000 feet, immediately inflate the life raft and life preservers. This allows adequate time for oral inflation if CO_2 charges fail.

References

Armstrong, H. G. (2017). Escape, Survival, and Rescue. In "Aerospace Medicine" (Chapter 20). Williams and Wilkins Co.

Carlson, A. J., Ivy, A. C., Krasno, L. R., & Andrews, A. H. (2018). Physiology of Free-Fall: Delayed Parachute Jumps. "Quarterly Bulletin of Northwestern University Medical School", 16, 254.

Cofer, F. S., Sweeney, H. M., & Frenier, C. E. (2020). "High-Speed Aircraft Escape Systems" (TSEAC-11-45341-1-2).

Colgan, J. W., & Hertzberg, H. T. E. (2022). "Pilot Prone Position Beds" (MCREXD-695-71D). Glazier, J. C., & Hallenbeck, G. A. (2019). "Free Fall Rates and Durations for Dummies and Humans" (TSEAD-696-100).

Haber, F. (2021). "High Altitude Escape and Survival" (USAF School of Aviation Medicine, Project No. 21-1207-0006).

Haber, F. (2020). Very High Altitude Bailout. "Journal of Aviation Medicine", 23, 322-329. Hallenbeck, G. A. (2018). Parachute Opening Shock Across Altitudes and Airspeeds. "MREng Div Report No. 49-696-66".

Johnson, L. F. Jr. (2025). "Current USAF Fighter Aircraft Escape Systems". (In Press).

Lovelace, R. W., & Allen, S. C. (2015). Parachute Descent from 40,000 ft Density Altitude (2015). "Eng-49-695-IV".

Lund, D. W. (2023). Positive G Effects on Centrifuge Subjects. "TSEA A-695-69".

Maison, G. L. (2016). "Dummy Descent Times with Various Parachute Types" (TSEAL 3-696-66H).

Mazza, V., Briggs, R. W., & Wheeler, R. V. (2020). "High Altitude Bailouts" (MREng Div No. 695-66M).

Shaw, R. S. (2021). "Short Duration Negative G Tolerance in Humans" (TSEA A-695-74). Shaw, R. S. (2017). "Human Acceleration Tolerance in Downward Seat Ejection" (TSEA A-695-74C).

Simmons, C. F. (2015). "P-1 Helmet and A-13A Oxygen Mask Wind Blast Visor" (AFTR-6037).

Smith, L. D. (2022). "Parachute Harness and Flight Clothing Integration" (MCREXD-666-19).

Strand, O. T. (2024). "Impact Performance of Two Protective Helmet Types" (AFTR-6020).

Sweeney, H. M., Savely, H. E., & Ames, W. H. (2016). "Ejection Seat Catapult Testing" (TSEA A-695-66H).

Wulff, V. J., Lovelace, W. R., & Baldes, E. J. (2019). "Ejection Seat for High-Speed Aircraft Emergency Escape" (MR-TSEAL-3-696-7aC).

Chapter 8: Pharmacological Considerations for Aircrew Members

This chapter focuses on the judicious use of pharmaceuticals for flying personnel. It emphasizes the Flight Surgeon's responsibility in managing and educating aircrew about medications that might impair flight performance. Specific attention is given to the dangers of self-medication and the effects of common drugs (e.g., antihistamines, nasal decongestants, antibiotics, anti-malarials) on flight safety and physiological responses like hypoxia. The chapter details drugs that can enhance or diminish hypoxia tolerance, as well as specific treatments for airsickness and fatigue. It concludes with an overview of Air Force policy on drug use for active flight duty and the particular risks associated with tranquilizers.

1. Flight Surgeon's Role in Drug Management

The administration of pharmaceuticals to flying personnel presents unique challenges and responsibilities, as few, if any, drugs are without potential implications for flight duty. The Flight Surgeon plays a pivotal role in ensuring aviation safety by meticulously managing drug use among aircrew members. This responsibility extends beyond merely prescribing medications; it encompasses diligent administrative control and comprehensive medical indoctrination of airmen. The primary objective is to prevent individuals from performing flying duties while under the influence of any medication that could impair their efficiency or compromise the safety of operations.

To effectively fulfill this role, the Flight Surgeon must remain thoroughly informed about all drugs, especially newly introduced ones. This constant vigilance is crucial to avoid prescribing any medication that might inadvertently jeopardize flight safety. A fundamental principle in this context is the acknowledgment of individual variability in drug susceptibility. What might be a benign dose for one individual could be detrimental to another, particularly within the demanding environment of flight.

The Flight Surgeon's duties also involve proactive medical education. Airmen must be made acutely aware of the specific dangers associated with using certain drugs while flying, as well as the significant risks inherent in self-medication.

This indoctrination should be continuous, emphasizing that flying personnel should only engage in flight duty when they are in optimal physical condition and entirely free from adverse pharmacological influences. This educational imperative extends to newly acclaimed drugs in popular media, particularly those with psychomotor or sensory effects such as depressants, powerful analeptics, antihistamines, and atropine-like compounds. By establishing a robust system of informed awareness and strict control, Flight Surgeons uphold the highest standards of safety for aircrew members.

2. Risks of Common Medications in Flight

A variety of commonly used medications, even those considered innocuous in terrestrial settings, can pose significant risks to flying personnel due to the unique physiological and operational demands of flight. The Flight Surgeon must ensure that aircrew members are thoroughly indoctrinated regarding these hazards, particularly concerning self-medication.

Antihistamines: Indiscriminate use of antihistaminics is especially dangerous. These drugs often exhibit a marked individual response, ranging from no noticeable effect in some individuals to pronounced drowsiness and even severe depression in others. Pharmacologically, antihistamines are known to depress the vestibular apparatus, which is critical for balance and spatial orientation. Furthermore, they can decrease depth perception. These combined effects create a substantial hazard for personnel attempting to fly while under their influence, directly compromising judgment and psychomotor functions essential for aviation.

Nasal Decongestants: While seemingly benign, nasal decongestants require careful administration. Due to the high absorptive capacity of the nasal mucosa, systemic effects can rapidly manifest. These may include tachycardia (increased heart rate) and various nervous states, such as tremors and incoordination, if used indiscriminately. Another potential hazardous side effect is mydriasis (dilation of pupils), which can impair vision and create a dangerous situation, especially during flight. Among this class of drugs, privine is noted as particularly dangerous in this context.

Prophylactic Drugs (Anti-malarials): Certain prophylactic medications, such as anti-malarials, can induce visual disturbances. Quinine and other drugs belonging to the cinchona group are known to cause tinnitus (ringing in the ears) and deafness, even in relatively small doses. These auditory effects can be individual responses to the drug and could interfere with communication and situational awareness in the cockpit.

Atropine-like Substances: Medications containing atropine-like compounds, such as hyoscine (a common ingredient in some cold remedies and airsickness pills) and banthine (used for ulcer symptoms), can lead to sufficient mydriasis and cycloplegia (paralysis of the ciliary muscle, affecting accommodation) to render them dangerous for flying personnel. Such visual impairments directly affect a pilot's ability to accurately perceive and react to environmental cues.

Antibiotics: The antibiotic group, while crucial in medicine, includes agents that can be very dangerous to flying personnel. Streptomycin and dihydrostreptomycin are particularly noted. Both can induce permanent hearing and vestibular damage. A critical distinction is that dihydrostreptomycin does not provide the warning symptom of dizziness that streptomycin often does, making its onset of vestibular damage potentially insidious. Therefore, it is strongly advised that these drugs not be administered to flying personnel if alternative treatments are available. Should their use be unavoidable, such as against H. Influenzae, thorough hearing and vestibular testing must be conducted both before initiating therapy and weekly throughout its duration. The drug should be immediately discontinued upon any detected decrease in function, unless it is absolutely essential for life-saving purposes. Chloromycetin poses another risk due to its adverse effects on the hemopoietic system, occasionally causing aplastic anemia. This impact on oxygen transport can decrease an airman's tolerance to hypoxia, necessitating careful control of its use in flying personnel.

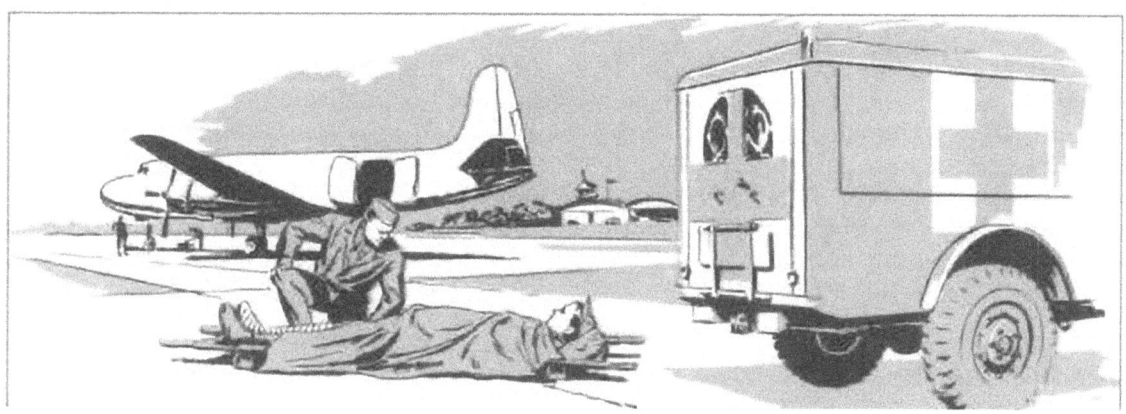

3. Drugs Affecting Hypoxia Tolerance

The physiological effects of hypoxia in the high-altitude flight environment are a critical concern, and certain pharmacological agents can either exacerbate or mitigate this risk. The actions of drugs in flight are often modified by hypoxia itself, alongside fatigue, underscoring the importance of understanding these interactions.

Drugs that Decrease Hypoxia Tolerance: Several common over-the-counter and prescription medications have been identified as potentially reducing an individual's tolerance to hypoxia. This reduction can stem from various mechanisms, primarily those that interfere with oxygen transport or utilization at the cellular level. * **Sulfonamides:** While some sulfonamides might not significantly impair hypoxia tolerance in all contexts, studies suggest that sulfanilamide, even in moderate doses, can diminish this tolerance. This effect is particularly noted in experiments where exacting or strenuous work is performed at sea level, indicating a potential for adverse impact under hypoxic conditions. Furthermore, abnormal toxic responses to sulfonamides, though occasional, could be significantly aggravated by the physiological demands of flight, making them more dangerous given the exacting nature of aviation tasks. If sulfonamide administration results in anemia or methemoglobinemia, or if the individual exhibits abnormal susceptibility, a reduction in hypoxia tolerance is certain. Sulfathiazole or sulfadiazine may adversely affect depth perception and phorias. * **Analgesics and Antipyretics:** Drugs such as APCs (Aspirin, Phenacetin, Caffeine), Bromo-Seltzer, and Alka-Seltzer, when consumed in large amounts or too frequently, can lead to the formation of methemoglobin. Acetilid and phenacetin, specifically, also induce methemoglobin formation. Methemoglobinemia reduces the blood's oxygen-carrying capacity, directly mimicking and thus worsening the effects of environmental hypoxia. This interference with oxygen transport is a direct threat to cognitive and physiological function at altitude.

Drugs that Increase Hypoxia Tolerance: Conversely, some drugs can improve an individual's ability to withstand hypoxic conditions, often by augmenting oxygen availability or reducing metabolic demand. * **Ammonium Chloride:** Administration of 10 to 20 grams of ammonium chloride per day for three days has been shown to increase arterial oxygen saturation by approximately 10% at 18,000 feet. This results in improved performance and a less pronounced acceleration of the pulse than would otherwise be expected under hypoxic conditions. The mechanism is believed to involve an increase in carbon dioxide exhalation, which subsequently increases alveolar oxygen tension. A subtle shift in the acid-base balance may also contribute by influencing oxyhemoglobin formation and dissociation. However, it is important to note that such doses of ammonium chloride frequently cause gastrointestinal irritation, which could be problematic in flight. * **Glucose:** Psychomotor and visual tests in humans suggest that glucose ingestion enhances performance at altitude. Evidence indicates that low blood sugar levels can impair the oxygenation of the central nervous system, meaning that even a mild oxygen deficit could produce symptoms that would not occur if blood sugar levels were normal. Additionally, a carbohydrate-rich diet leads to a higher alveolar respiratory quotient, which plays a role in decreasing alveolar carbon dioxide tension. These physiological benefits support the rationale for consuming carbohydrate-rich foods immediately prior to high-altitude missions. * **Analeptics:** Under hypoxic conditions, analeptics such as benzedrine (10 mg), pervitin (methedrine, 5 mg), dexedrine (5 mg), or caffeine sodium benzoate (500 mg) can improve psychomotor performance. There is some evidence suggesting benzedrine's superiority over caffeine for this purpose. However, a significant body of evidence also indicates that therapeutic doses of benzedrine or dexedrine can adversely affect judgment. Consequently, extreme caution must be exercised when these drugs are administered to flying personnel, and their use should be carefully monitored due to the potential for impaired decision-making critical for flight safety.

4. Medications for Airsickness and Fatigue

Airsickness and fatigue are two significant challenges for aircrew members, particularly during prolonged missions or in the early stages of training. Pharmacological interventions exist for both, but their use in flying personnel requires careful consideration due to potential side effects that could compromise flight safety.

Medications for Airsickness: Airsickness medications typically fall into three broad categories: parasympathetic depressants, central nervous system depressants, and antihistaminics. While some drugs are effective across various types of motion sickness, their suitability for aircrew hinges on their ability to prevent sickness without impairing performance or causing undesirable side effects. Ideal anti-airsickness drugs should be non-toxic, non-habit forming, free from disagreeable symptoms, and rapidly active after oral administration.

- **Barbiturates:** Central nervous system depressants like barbiturates are generally of doubtful value for flying personnel. Studies on swing sickness and seasickness have not consistently demonstrated their effectiveness, nor do they significantly enhance the beneficial effects of other motion sickness drugs. Phenobarbital, barbital, amytal, and pentobarbital have been among those studied. However, it is noted that the Army's motion sickness preventive, which combines sodium amytal with hyoscine and atropine, has shown efficacy in various motion sickness contexts, though it is unclear if this is solely due to the hyoscine and atropine components. The inherent CNS depression associated with barbiturates makes them unsuitable for active flight duty.

- **Atropine Series (Scopolamine):** Among the numerous compounds and mixtures tested, members of the atropine series have consistently provided substantial protection against airsickness. Scopolamine, in particular, has been the most widely used anti-motion sickness drug due to its high degree of protection and relatively low incidence of side effects. However, atropine-like substances can cause mydriasis and cycloplegia, impairing vision.

- **Antihistamines:** The report by Gay and Carliner in 2023 on dramamine's effectiveness in seasickness spurred significant interest in antihistamines for motion sickness. While initial tests showed scopolamine to be more effective than dramamine or benadryl, subsequent research explored combinations. A mixture of hyoscine hydrobromide (0.65 mg) and benadryl (50 mg) demonstrated greater protection than scopolamine alone, with a half-dose preparation proving as effective as full-dose hyoscine but with fewer side effects. Other preparations noted for good protection include phenergan, trimeton, marezine, pyrrolazote, and bonamine. The School of Aerospace Medicine's extensive testing program indicates little statistical difference in effectiveness among conventional drugs like marezine, dramamine, bonamine, and phenergan. The optimal choice often depends on individual patient response.

- **Restrictions for Aircrew:** Crucially, none of these anti-motion sickness drugs have been demonstrated as safe for use by active aircrew members. Their use is strictly limited, on occasion, to cadets and officers in early training periods, and only when the student is not in primary control of the aircraft (e.g., during dual instruction). Such use must be under the careful control of the Flight Surgeon, who is responsible for preliminary testing to detect unusual hypersensitivity reactions and for informing and indoctrinating the instructor pilot. The duration of use should be short, typically three to four weeks, to support the individual in developing natural resistance to motion effects, rather than fostering dependence. The Flight Surgeon must rigorously enforce control to prevent unauthorized self-medication. Drugs containing belladonna alkaloids are specifically mentioned to paralyze accommodation and impair vision in therapeutic doses, highlighting a direct conflict with flight requirements.

Medications for Fatigue: Fatigue is a significant factor in mission performance, and certain drugs can temporarily postpone its effects and maintain alertness. * **Dexedrine (Dextro-amphetamine):** Dexedrine allows individuals to postpone sleep and fatigue, maintaining alertness for many hours beyond normal limits. The typical dosage is 2.5 mg of dexedrine sulfate every three hours, or 5 mg every six hours. This drug does not significantly impair physical or mental efficiency in well-integrated subjects; rather, it enables them to continue duties for extended periods. While not habit-forming in the physical sense, its stimulant effects can be perceived as pleasant, leading to potential excessive use. Overdosage can cause excitement, headache, and sleeplessness. Due to individual susceptibility, personnel should be tested with a 2.5 mg dose before combat use. It is critical to understand that dexedrine is not a substitute for rest or sleep; it merely postpones the physiological need for them.

5. Air Force Policy on Drug Use for Flying Personnel

The United States Air Force maintains a stringent policy regarding drug use by flying personnel, emphasizing aviation safety as paramount. While the Surgeon General respects a physician's prerogative to practice sound medicine, Air Force policy generally dictates that no drugs are to be used by an individual while on flying status. This broad directive underscores the inherent risks that even seemingly minor medications can pose in the complex and demanding aviation environment.

A primary responsibility of the Flight Surgeon is to actively detect and prevent self-medication among aircrew members. This requires vigilant observation and proactive measures to ensure that flying personnel do not obtain or use drugs either independently or through other allied medical service personnel who are not Flight Surgeons. To effectively manage this, Flight Surgeons must cultivate strong rapport and liaison with other healthcare providers, including ward officers and dentists. This collaboration ensures that the Flight Surgeon is informed whenever flying personnel are receiving treatment, allowing for appropriate assessment of flight status.

In the rare and highly controlled instances where analeptics, such as dextro-amphetamine, are utilized, the Flight Surgeon must exercise the utmost control. These stimulants are explicitly not to be used routinely or as substitutes for essential personal hygiene, such as adequate rest. Their application is restricted to situations where a decrement in performance is imminent, and only after thorough briefing and pretesting for any unusual individual responses. This cautious approach ensures that such drugs are employed solely for critical operational needs, with minimal risk.

Furthermore, the use of several classes of drugs by individuals on flying duty is contraindicated due to considerable associated risks. These include: * **Antihypertensives:** Medications used to manage high blood pressure can have systemic effects that might compromise a flyer's physiological response to gravitational forces or other stressors. * **Anticholinergics:** These drugs can cause visual disturbances (e.g., mydriasis, cycloplegia) and cognitive impairments, directly affecting critical flight functions. * **Antihistaminics:** As previously discussed, their sedative and vestibular depressant effects pose significant hazards to depth perception and spatial orientation. * **Sedatives:** Drugs that induce drowsiness or reduce central nervous system activity are fundamentally incompatible with the alertness and cognitive function required for flight.

The mere indication for using these drugs, or other similar medications, is generally considered sufficient cause for an individual's removal from flying status. This strict policy reflects a commitment to ensuring that aircrew members are unencumbered by pharmacological effects that could impair their judgment, reaction time, or physiological resilience, thereby safeguarding both individual lives and mission success.

6. Tranquilizers and Flight Safety

The use of tranquilizers by aircrew members presents significant safety concerns, directly impacting an individual's ability to cope with acute physical and psychological stress, which are inherent elements of flight operations. Research into the effects of tranquilizers on stress tolerance has provided clear recommendations regarding their use in flying personnel.

A study conducted at the Aero-Medical Laboratory, WADD, on healthy male volunteers, evaluated tolerance to acute physical stress while subjects were under the influence of tranquilizers. The study used subjects as their own controls, comparing their performance with and without medication. The findings revealed that stress tolerance was significantly reduced when individuals were taking tranquilizers, specifically promazine hydrochloride and meprobromate. As the intensity of physical stress increased, the volunteers exhibited a severe limitation in their response capabilities while influenced by both types of tranquilizing agents.

In addition to these physiological limitations, psychological changes were also observed in the subjects. These alterations in cognitive and emotional states further underscore the incompatibility of tranquilizer use with the high-stakes environment of aviation.

Based on these findings, a strong recommendation was made: if an individual's medical condition necessitates the use of tranquilizers, that aircrew member should be removed from flying status for the entire duration of the medication. This policy prioritizes flight safety by preventing any potential impairment of an airman's ability to perform critical duties under stressful conditions, thereby minimizing risks to themselves, their crew, and the mission.

References

Ashe, W. F. (2020). Drugs: Friend or Foe for Aviators? In "Proc. Aviation Medicine Symposium on Toxic Hazards in Military Flying and the Industry". Headquarters AMC.

Armstrong, H. G. (2018). Aircrew Maintenance. In "Aerospace Medicine" (Chapter 26). Williams and Wilkins Co.

USAF School of Aviation Medicine. (2015). "Chloroquine Effects on Visual Accommodation Speed" (Project No. 21 1601-005).

Gay, L. N., & Carliner, P. E. (2017). Motion Sickness Prevention and Treatment: Seasickness. "Science", 109, 359.

Glorig, A. (2021). Dihydrostreptomycin's Impact on the Auditory System. "Annals of Otology, Rhinology & Laryngology", 60, 327-335.

Greenwood, G. J. (2019). Neomycin-Induced Ototoxicity. "Archives of Otolaryngology", 69, 390-397.

Haas, W. R. (2016). Antimalarials in Flight Personnel. "Journal of Aviation Medicine".

Leake, C. D. (2020). Amphetamines and Driver Alertness. "Ohio Medical Journal", 53, 176-178.

Lett, J. E. (2015). Nasal Vasoconstriction in Aviation. "USAF Medical Service Digest", III(5). Naunton, R. F., & Ward, P. H. (2017). Kanamycin Ototoxicity with Renal Impairment.

"Archives of Otolaryngology", 69, 398-399.

Smith, G. M., & Beecher, H. K. (2019). Amphetamine, Secobarbital, and Athletic Performance: Subjective Evaluations. "Journal of the American Medical Association", 172, 1502-1514.

Smith, G. M., & Beecher, H. K. (2020). Amphetamines, Secobarbital, and Athletic Performance: Effects on Judgment. "Journal of the American Medical Association", 172, 1623-1633.

Shambaugh, G. E. Jr. (2022). Dihydrostreptomycin-Induced Hearing Loss. "Journal of the American Medical Association", 170, 1657-1660.

Waters, R. O. (2015). "Ototoxic Medications" (USAF Aerospace Medical Center, School of Aviation Medicine Review 5-60).

Chapter 9: Fatigue Management in Aerospace Operations

The pervasive challenge of fatigue in aerospace operations represents a critical concern, impacting both combat readiness and overall safety. This chapter undertakes a comprehensive examination of fatigue, defining its various forms within the unique context of aviation and space endeavors. It delves into the distinctions between physiological fatigue, characterized by metabolic changes, and "skill fatigue," which emerges from the cognitive and psychological demands of prolonged operational tasks. The discussion further explores the observable manifestations of skill fatigue, identifies its numerous contributing factors, and introduces the concept of "chronic flying fatigue" as a cumulative state resulting from inadequate recovery periods between missions. Emphasis is placed on the intricate interplay of environmental and personal factors that influence an individual's susceptibility to fatigue. Finally, the chapter outlines robust management and prevention strategies, underscoring the indispensable role of human factors engineering, leadership, and proactive crew control in mitigating fatigue's detrimental effects, while also considering the evolving challenges posed by future long-duration space missions.

Definition and Impact of Fatigue in Aerospace Operations

Fatigue, a term derived from the Latin fatigare, meaning "to waste away," is a multifaceted phenomenon whose interpretation varies across scientific and technical disciplines. While metallurgists may employ the term to describe the progressive deterioration of structural strength under repeated stress, and physiologists utilize it to characterize decrements in organic responses stemming from sustained activity, the context of aerospace operations necessitates a more specialized understanding. For the Flight Surgeon, operating at the vanguard of aeromedical practice, fatigue is most appropriately defined as a "detrimental alteration or decrease in skilled performance related to the duration or repetitive use of that skill, aggravated by physical, physiological, and psychic stress." This definition encapsulates the complex interplay of factors inherent in the aerospace environment, highlighting its direct relevance to the operational capabilities of aircrews.

The implications of fatigue in aerospace are far-reaching and critically significant. It constitutes an ever-present threat to the efficacy and safety of military and civilian aviation alike. The contemporary operational landscape, characterized by global strategic and tactical capabilities, relies heavily on the ability to traverse vast distances, often without the luxury of established airfields. The advent of in-flight refueling techniques, for instance, has enabled aircraft initially designed for relatively brief missions to remain airborne for extended periods. This progression in aircraft performance and complexity, coupled with significantly increased flight times, has, to some extent, outpaced advancements in aeromedical science aimed at ensuring equivalent aircrew capability.

Consequently, Flight Surgeons consistently encounter fatigue-related issues, irrespective of their unit's specific mission profile. Whether it involves managing crews engaged in 24-hour B-52 flights, overseeing overseas tactical deployments in fighter aircraft, supporting around-the-clock troop carrier and Military Air Transport Service (MATS) operations across disparate geographical regions, or meticulously planning the daily routines of astronauts, fatigue remains a central challenge. The impact extends beyond mere operational efficiency; tragically, fatigue demonstrably contributes to aviation accidents, exacting a steady toll on human lives.

Evidence from the Directorate of Flight and Missile Safety Research strongly suggests that many erroneous decisions leading to aircraft losses, particularly at the culmination of protracted missions, can be attributed to the insidious effects of fatigue. Despite the undeniable gravity of this problem, the Flight Surgeon's ability to quantitatively measure fatigue in any single individual remains severely constrained. Numerous attempts to develop an operationally applicable fatigue index have, to date, been unsuccessful. While many of the physical, physiological, psychological, and environmental factors known to contribute to fatigue can be identified and analyzed, the underlying mechanisms governing this complex phenomenon are still only partially understood. This gap in understanding underscores the necessity for ongoing research and a multifaceted approach to fatigue management in aerospace.

Physiological and Skill Fatigue Explained

To comprehensively address fatigue within the aerospace domain, it is essential to distinguish between its physiological underpinnings and the more operationally relevant concept of "skill fatigue." Traditional physiological perspectives, often rooted in kymographic studies of repeatedly stimulated muscle-nerve preparations, have long associated fatigue with specific metabolic alterations. Physiologists have observed that sustained muscular activity leads to a depletion of energy resources, concurrently with a rapid accumulation of anaerobic breakdown products, such as lactic acid. This process typically precedes the observable loss of muscle irritability. Oxygen's fundamental role in mitigating metabolite accumulation through aerobic oxidation within the tricarboxylic acid cycle further highlights the biochemical nature of this type of fatigue.

In the context of the exercising human, these metabolic principles hold true: vigorous physical effort results in an increased energy turnover and a buildup of anaerobic products, commonly manifested as an "oxygen debt" following exertion. Such observations have historically led physiologists to conceptualize fatigue as an energy-depleted state inextricably linked to metabolite accumulation. From this viewpoint, contractile decrement is merely a symptom of a broader metabolic complex. This conceptual framework has frequently been extrapolated to suggest that a wide array of biological regression phenomena are ultimately attributable to a fundamental organic state wherein cells, whether muscular or neural, lose their inherent capacity to respond effectively to stimulation.

However, applying these physiologically derived concepts directly to ergographic observations in intact organisms, particularly humans engaged in complex tasks, presents inherent difficulties. For instance, a human subject can engage in manual work to the point of perceived exhaustion, yet electrical stimuli applied near the motor-nerve end-plate can still elicit a contraction comparable to that of a rested individual. This demonstrates that the decrement observed in the intact organism is often far removed from the crippling biochemical changes evident in a constantly stimulated, isolated muscle-nerve preparation. A pertinent example illustrates this paradox: a subject may write symbols for hours until their work degenerates into illegible scrawls; however, if subsequently asked to write their name, they perform with the same proficiency as if entirely rested. This observation clearly indicates that an observed reduction in output does not necessarily signify a commensurate decrease in underlying capacity. Some physiologists refer to this nuanced distinction as "central fatigue," implying a fatigue that resides not in the effector organs but within the central nervous system's ability to initiate and sustain volitional effort. Importantly, the human subject enduring repetitive, intensive, and prolonged mental stress typically does not exhibit the same biochemical changes observed in muscles affected by the physiologist's traditional definition of fatigue.

For the Flight Surgeon, the primary objective is to identify the precise reasons for performance deterioration and to judiciously manipulate etiological circumstances to delay the onset and mitigate the progression of fatigue. In achieving this, the Flight Surgeon must often contend with factors and conditions that extend beyond conventional notions of physiological fatigue. Within the demanding context of the flying situation, fatigue is generally categorized into two primary, often overlapping, forms: acute skill fatigue and chronic fatigue. Both categories demand the Flight Surgeon's meticulous attention to the flyer's ground and air environment, as they represent distinct yet interconnected challenges to operational effectiveness and safety.

Manifestations and Contributing Factors of Skill Fatigue

Acute single mission skill fatigue represents a form of performance decrement directly attributable to the wearing repetition of tasks during an extended mission or a series of successive short missions. This type of fatigue is a common experience among individuals engaged in mentally and physically demanding activities, and crucially, it typically resolves with adequate rest. Its characteristic symptoms include generalized lassitude and a discernible disinclination to engage in further activity. The inherent nature of flying, which demands prolonged and detailed attention to a multitude of concurrent tasks, renders the emergence of skill fatigue a natural and expected phenomenon.

The specific etiological pattern of skill fatigue is highly individual, varying considerably among aircrew members. The Flight Surgeon, leveraging their own flying experience, is uniquely positioned to evaluate the various extrinsic factors that contribute to this condition. These external stressors encompass a range of environmental and operational elements: * **Hypoxia:** Reduced oxygen availability, even at sub-symptomatic levels, can subtly degrade cognitive function and accelerate fatigue onset. * **Temperature Extremes:** Both excessively hot and cold conditions impose physiological strain, diverting metabolic resources and accelerating the perception of fatigue. * **Noise:** Prolonged exposure to high-intensity aircraft noise, particularly within cockpits, can contribute significantly to auditory fatigue and a generalized sense of exhaustion. * **Vibration:** Constant low-frequency vibrations, endemic to aircraft environments, induce physical discomfort and sensory overload, contributing to weariness. * **G-forces:**

Sustained or repeated exposure to accelerative forces can induce physiological stress and muscle strain, further exacerbating fatigue. * **Inability to Move About:** Confinement within cramped cockpits for extended periods leads to physical stiffness, reduced circulation, and psychological discomfort. * **Feeding and Elimination Problems:** Irregular or inadequate access to nutrition and facilities for elimination can disrupt physiological rhythms and negatively impact morale. * **Other Considerations Inherent in Most Aircraft:** This broad category encompasses a myriad of specific aircraft design limitations or operational quirks that contribute to discomfort and stress.

In addition to these extrinsic factors, Flight Surgeons must also consider intrinsic psychological elements that profoundly influence skill fatigue: * **Boredom:** Particularly relevant during monotonous phases of a mission, boredom can lead to reduced vigilance and increased susceptibility to errors. * Responsibility: The immense responsibility associated with flying complex aircraft and safeguarding human lives or high-value assets imposes significant cognitive load. * **Frustrations:** Technical malfunctions, communication issues, or procedural delays can generate frustration, consuming mental energy. * **Problems of Attention and Concentration:** The demanding requirement for sustained attention across multiple information streams and the need for intense concentration on critical tasks are highly fatiguing. * **Apprehension, Anxiety, and Uncertainty:** These emotional states, stemming from the inherent risks of aviation, contribute to a baseline level of psychological stress. * **Fear:** Whether overt or operating at a subconscious level, the realization of potential hazards—such as fire, engine failure, emergency ejection over water, or complex low-altitude escape scenarios—creates a persistent undercurrent of apprehension. This "subconscious feeling" can establish a baseline of anxiety that contributes significantly to the gradual erosion of a flyer's energetic reserves.

Studies utilizing simulated airborne performance environments, such as the Cambridge cockpit, have delineated a clear set of skill fatigue components and associated subjective phenomena. These manifestations include: **1. Increased Stimuli Requirement:** A noticeable need for larger than normal stimuli to elicit appropriate and timely responses. **2. Timing Errors:** An increase in errors related to the precise timing of actions and decisions. **3. Element Overlooking:** A propensity to miss or overlook important elements within a sequential task series. **4. Control Impairment:** A distinct loss in the accuracy and smoothness of control column and rudder movements. **5. Unawareness of Errors:** A concerning unawareness of the accumulation of rather significant errors in critical parameters such as azimuth, elevation, and aircraft attitude. **6. Compensatory Control Efforts:** An increase in control movements, often involving greater fluctuations, necessary to produce the same desired effect, indicating reduced efficiency. **7. Under- and Over-Control:** The tendency to alternately under-compensate or over-compensate during control inputs. **8. Side-Task Neglect:** Forgetting or neglecting secondary, albeit often important, tasks. **9. Unreliable Reporting:** An increasing unreliability in reports concerning transpired events, reflecting memory and perception deficits. **10. Errors of Inattention:** Manifestations such as reduced visual scanning, leading to "fixed vision" and a failure to adequately monitor the surrounding environment or instrument panel. **11. Task Preoccupation:** A tendency to become excessively focused on one component of a task to the exclusion of others. **12. Operational Sequence Disruption:** Allowing various elements of the operational sequence to appear out of synchronization or place with respect to one another.

Concurrently with these objective performance indicators, various subjective phenomena are frequently reported: **1. Increased Physical Discomfort:** A heightened awareness and intolerance of physical discomforts. **2. Growing Irritability:** An escalating sense of irritability, both internally and externally manifested. **3. Projection of Irritability:** A tendency to project this irritability onto aspects of the machine or immediate environment. **4. Unawareness of Deficiencies:** A paradoxical increasing unawareness of one's own performance deficiencies. **5. Other Tension and Unpleasantries:** A general sensation of heightened tension and psychological unpleasantness.

In essence, the perceptual-motor aspects of the task, under the influence of skill fatigue, tend to disintegrate from an organized, integrated whole into a diversity of unrelated stimulus elements, leading to corresponding dissociations of action and a loss in "central" cognitive control. Crucially, these signs rarely indicate an actual loss in the *ability* to perform; rather, they suggest that actions are not performed effectively unless an appropriate and substantial special effort is consciously expended.

The maintenance of a high level of efficiency in aircrew function is absolutely vital for ensuring combat effectiveness and paramount for flying safety. Defects arising from skill fatigue, such as reduced visual scanning, are particularly dangerous in today's crowded airspace, significantly increasing the probability of mid-air collisions. Similarly, decrements in judgment and attentiveness contribute directly to incidents of disorientation and loss of aircraft control, issues frequently observed at low altitudes. A particularly insidious aspect of skill fatigue is that even when individual flyers recognize that they are tired, they may remain unaware of, or be unwilling or unable to admit, the extent of their performance deterioration.

Other specific fatigue factors depend heavily on the individual aircrewman's particular tasks. For instance, visual fatigue can be a prominent concern for radar scanners or personnel constantly working with charts. For others, such as an isolated tail gunner, the sheer monotony and boredom of the task can become overwhelmingly fatiguing. Across all crew positions, overt muscular effort might be minimal, but the pilot's actions are fundamentally rooted in making highly skilled responses to signals within their immediate environment, primarily derived from a continuous stream of instrument readings. This involves a ceaseless interplay between the interpretation of complex signal groups and the execution of precisely timed and coordinated responses. Through rigorous training and extensive experience, pilots develop the capacity to anticipate and prepare for successive responses based on preceding cues, thereby ensuring smooth, safe, and efficient flight operations. However, skill fatigue directly undermines this critical anticipatory capability.

Conceptualizing the balance between fatigue and alertness involves three key principles:
1. Threshold of Indifference: In a rested state, a minimal detectable change elicits a minimal response. With the onset of fatigue, the mind and body become indifferent to these subtle cues, failing to react appropriately to changes that would normally be readily perceived.
2. Anticipation Span: Flying is fundamentally an operation meticulously laid out as a sequence of closely related steps. Pilots rely heavily on anticipation to remain abreast of this sequence and maintain proactive control. Fatigue, however, causes a contraction of this crucial anticipation span. This leads to the emergence of forced and hurried reactions, disrupting the pilot's smooth rhythm and often resulting in potential emergencies being addressed too late. **3. Speed and Load:** These factors have gained immense importance in the era of high-speed aircraft. The sheer speed at which modern aircraft operate is so great that flight control systems must provide advanced information to pilots. Furthermore, the operational load is so high that sophisticated computer systems have become standard equipment, effectively offloading many complex control functions from the pilot and integrating vital information in ways that enhance situational awareness without overwhelming the human operator. However, this increased speed, combined with a narrower margin of control lead time and an escalating mass of simultaneous stimuli, collectively generates tension, elevates the intensity of work, demands greater alertness, and inexorably accelerates the tempo towards fatigue.

Empirical observations from military aviation underscore these points. The Royal Air Force (RAF), for example, documented objective and measurable fatigue in jet fighter pilots after as few as three consecutive one-hour sorties. Similarly, fatigue was evident after 10 hours in multi-engine piston aircraft and after 6 hours in jet bombers. This fatigue was found to be more pronounced during night flying operations and progressively worsened during four consecutive 15-hour night missions. Interestingly, it was often present, though not subjectively acknowledged, after three fighter sorties, and its severity varied widely among individual pilots operating under identical circumstances. These findings continually drive human factors engineers to seek ongoing improvements in cockpit design, develop advanced systems for feeding computer-digested information as tactile "feel" to controls, simplify instrument data presentation, and enhance radio-navigational and automatic pilot systems.

Bartley and Chute offer a psychological perspective, viewing fatigue as an experimental pattern arising from a conflict situation. In this framework, the individual develops an aversion towards the task they are currently performing or are about to undertake. This perspective suggests that feelings of unwillingness, inadequacy, and frustration coalesce to manifest as fatigue—an implicit attempt to retreat from or escape a task or situation that is perceived as becoming increasingly difficult to manage. They characterize feelings of fatigue and boredom as avoidance habits, reinforced by their effectiveness in reducing psychological tension. While these are typically mild and universal phenomena in everyday life, under extreme conditions, they can escalate to the neurotic proportions characteristic of "battle-fatigue," "operational-fatigue," and other comparably vivid combat syndromes.

In the context of single mission fatigue within the usual flying situation, the Flight Surgeon's primary concern may not necessarily delve into the intricate psychological dynamics of aversion. Regardless of whether fatigue represents an avoidance reaction, spending 24 hours in an ejection seat is undeniably a protracted period, and fatigue is the natural, normal response to such sustained exertion and confinement. The Flight Surgeon's overarching objective, therefore, is to minimize the onset and severity of this fatigue and to ensure it remains within controllable limits.

Chronic Flying Fatigue: Causes and Symptoms

Chronic flying fatigue represents a cumulative phenomenon that arises when the physical and mental recuperation between repeated missions is incomplete. Unlike acute skill fatigue, which resolves with a single period of adequate rest, chronic fatigue builds progressively over time, signifying a more profound and persistent state of weariness. Its development is a complex interplay between two primary sets of factors: firstly, the inherent difficulty, frequency, and duration of the missions undertaken; and secondly, the adequacy, duration, and effectiveness of the time allocated for rehabilitation between these missions. This insidious form of fatigue can manifest rapidly, sometimes within a mere one or two weeks, particularly in the context of repetitive, maximum-effort mission programs.

Historical accounts from military operations provide vivid illustrations of chronic flying fatigue. Royal Air Force (RAF) Flight Surgeons, during the intense period of the Berlin Airlift, observed that a striking 90% of their aircrew members exhibited symptoms of fatigue early in the operation. These symptoms were wide-ranging and included pervasive tiredness, heightened apprehension, an increase in alcohol consumption, noticeable weight loss, interpersonal bickering, and a multitude of minor physical complaints that often lacked clear organic bases. Concurrently, commanders began to observe distinct deteriorations in operational performance. These included bumpier landings, careless taxiing procedures, clumsy handling of aircraft controls, and a marked increase in careless flight planning. These performance decrements often appeared to worsen later in the day, after crews had completed several missions. In some instances, pilots even exhibited alarming lapses, such as forgetting to feather engines or deploy landing flaps, behaviors indicative of severe cognitive overload and exhaustion. Crew questionnaires administered during this period revealed that a significant portion of these difficulties stemmed from factors such as inadequate or disordered sleep, prolonged waiting periods between flights, substandard living conditions, excessively long duty hours, deficiencies in ground and air organizational support, poor food quality, and in-flight discomforts.

Crucially, the incidence of fatigue decreased markedly when vigorous and proactive attention was directed toward improving these environmental factors. Measures such as relocating sleeping areas away from the noisy flight line, publishing duty schedules well in advance with clearly programmed off-duty time, enhancing food quality, and relieving pilots of burdensome administrative details proved highly effective in mitigating chronic fatigue.

A similar study conducted by the Royal Canadian Air Force (RCAF) during the Tokyo Airlift further corroborated these findings, identifying irritability, pervasive sleepiness, a noticeable lowering of performance standards, and delayed reaction times as the chief manifestations of flying fatigue. This study meticulously categorized the relevant factors into three main groups:

1. **Factors Common to Flying in General:**
   - **Length of Flight:** Extended flight durations inherently increase physical and mental strain.
   - **Delayed Flights:** Unforeseen delays can disrupt pre-planned rest schedules and create additional stress.
   - **Details Prior to Takeoff:** Extensive pre-flight preparations and briefings contribute to cumulative workload.
   - **Reliability of Radio Communication and Navigational Aids:** Poor system reliability increases cognitive load and anxiety.
   - **Monotony and Boredom on Familiar Routes:** Repetitive tasks over familiar routes can lead to reduced vigilance.
   - **Number of Intermediate Stops:** Frequent stops disrupt rest, prolong duty periods, and add logistical complexity.
   - **Drinking the Night Before:** Alcohol consumption, even if not leading to overt intoxication, can profoundly impair subsequent performance and recovery.

2. **Factors Relating Specifically to Operational Conditions:**
   - **Problems Related to Particular Aircraft:**
     - **High Levels of Noise and Vibration:** These can induce significant physiological stress and auditory fatigue.
     - **Unreliable and Inadequate Heating Systems:** Thermal discomfort compromises crew well-being and operational focus.
     - **Cramped Working Conditions, Especially in the Cockpit:** Physical confinement contributes to discomfort, stiffness, and psychological strain.
     - **Poor Arrangement of Instruments:** Suboptimal cockpit ergonomics increase cognitive load and the likelihood of errors.
     - **Uncomfortable Oxygen Masks:** Ill-fitting or uncomfortable masks add to physical stress and distraction.
   - **Problems Relating to Particular Route:**
     - **Food:** Inadequate or unpalatable food on specific routes impacts nutrition and morale.
     - **Quarters:** Poor quality or noisy accommodation prevents effective rest and recovery.
     - **Transportation:** Difficult or unreliable transport to and from duty stations adds to physical and temporal strain.
     - **Hours:** Unpredictable or excessively long duty hours disrupt circadian rhythms and sleep.

3. **Personal Factors:**
   - **Inexperience:** Novice aircrews may exhibit lower stress tolerance and be more susceptible to fatigue.

- o **Tension Among Crews:** Interpersonal conflicts or lack of cohesion within a crew exacerbates psychological stress.
- o **Responsibility:** The weight of operational responsibility can be a significant psychological burden.
- o **Domestic Worries:** Personal and family issues can divert mental resources and reduce resilience to operational stressors.
- o **Personalities:** Individual personality traits, including coping mechanisms and emotional resilience, profoundly influence fatigue susceptibility.

These multifaceted factors are applicable to virtually all aerospace operations where flying fatigue is a significant concern. For example, a scenario involving a reduction in MATS crew strength, forcing fewer personnel to handle the same transport load, quickly revealed the onset of fatigue. This situation was exacerbated by simultaneous requirements to support a massive maneuver, respond to a major disaster in Chile, and manage a crisis in the Congo. Interestingly, while "static" skill fatigue was observed during arduous 18-hour flights in tropical environments, chronic fatigue remained minimal among these crews. This outcome was primarily attributed to their high motivation and the tangible sense of accomplishment derived from their efforts—a certain number of tons delivered to destinations where the need was acute. Despite improvised facilities, crews were able to rest effectively, and the food provided was generally good, reinforcing the importance of motivational and supportive environmental factors.

Conversely, in the Continental United States, pilots engaged in routine operations, already flying at near maximum capacity, were suddenly compelled to compensate for the prolonged absences of other crews. These pilots were subjected to the insidious boredom of working longer hours on the same repetitive tasks, lacking the stimulating urgency of a new and special mission. Despite intensive efforts, maintaining motivation and a sense of accomplishment proved challenging. Many individuals found themselves flying over 125 hours per month, often compounded by extensive layovers. Furthermore, many were away from home for 25 or more days each month, only to return and immediately face alert duties or other routine assignments.

This relentless operational tempo profoundly disturbed diurnal sleep rhythms. For instance, crews returning from night flights would arrive home during daytime hours, when children were noisily active. With crew rest periods often concluding within 12 to 15 hours, these personnel were frequently compelled to resume flying duties after having achieved very little genuine rest. After just a few months of such a schedule, Flight Surgeons began to observe numerous indicators of chronic fatigue. In flight, aircrews exhibited increased complacency, accepting aircraft for the homeward leg of a trip that they would never have considered flying away from home. Concurrently, there was a noticeable increase in accident rates.

At sick call, aircrew members presented to the Flight Surgeon with complaints of colds and sore throats, frequently with an absence of clear physical findings. Further questioning by the Flight Surgeon often elicited a consistent narrative of excessive flying, profound fatigue, and an overt desire for more time on the ground. The incidence of tension-related illnesses, such as neurodermatitis, tension headaches, and peptic ulcers, also saw a marked increase.

A particularly acute situation was documented at a staging base for Congo operations, where Flight Surgeons encountered numerous aircrewmen who had logged 100 hours of flight time in less than ten days. Many others had accumulated up to 180 hours per month and were observed falling asleep during the brief intervals between laboratory tests in the clinic. A significant number exhibited resting pulse rates exceeding 100 beats per minute, which did not decrease appreciably even during sleep. Elevated blood pressures were also common, and universally, these individuals presented as profoundly tired. Such combinations of severe chronic fatigue, superimposed on a series of exhausting long missions, and exacerbated by skill fatigue under the pressure of urgent operational demands, presented formidable challenges in clinical management for the Flight Surgeons.

Fatigue Management and Prevention Strategies

The effective management of fatigue in aerospace operations is fundamentally predicated on prevention, largely due to the inherent difficulty in its objective measurement. The absence of a "red warning light" in the cockpit to signal fatigue, coupled with the Flight Surgeon's lack of a calibrated device for quantitative fatigue assessment, renders proactive strategies paramount. When preventative measures prove insufficient, whether due to individual factors, the inherent nature of the mission, or deficiencies in support facilities, diagnostic recognition becomes the subsequent critical step. Consequently, prevention and management efforts must encompass comprehensive pre-flight, in-flight, and post-flight control of both the aircrew member and their operational environment. This necessitates a collaborative "team effort" involving the individual flyer, their commanding officers, the Flight Surgeon, and the engineers responsible for designing aircraft and weapons systems.

Significant progress in mitigating fatigue has already been achieved through advancements in human factors engineering. Unlike earlier periods, the human operator is no longer treated as a last-minute addition to aircraft design. Modern cockpits are designed to be more spacious, and personal equipment is continuously refined to enhance comfort and range of motion. Instrument displays are now simpler and more intuitive, thereby reducing the likelihood of errors and eliminating time and energy-consuming steps in data interpretation. Furthermore, sophisticated computer systems have become standard, offloading many complex control functions from the pilot and integrating vital information in ways that enhance situational awareness without overwhelming the human operator.

The role of command and leadership is pivotal in fostering mission orientation and motivation among aircrews. It is the commanding officer who ultimately determines mission planning, allocates crew loads, and ensures the adequacy of support facilities. This leadership involves a delicate balance, weighing the risks of fatigue and potential accidents against the critical importance of the mission, available funding, and even the personal rewards associated with accomplishing what might appear to be an almost impossible mission load.

The contributions of all echelons of support personnel are undeniably crucial, a fact vividly underscored by the findings from the Tokyo and Berlin Airlift reports. These historical analyses demonstrated that seemingly peripheral factors, such as the quality of accommodation, food, and logistical support, profoundly impact aircrew morale and resilience to fatigue. The "facets" of a robust aircrew effectiveness program, encompassing physical, physiological, and psychological well-being, naturally form the cornerstones of a comprehensive fatigue prevention program. To reduce the vulnerability of the human component to fatigue, a structured approach involving post-flight, pre-flight, and in-flight crew control is an invaluable conceptual framework.

The prevention of future fatigue begins with the successful conclusion of the current mission. The primary objective of post-flight crew control is the effective utilization of recuperative time between missions. The Flight Surgeon plays a key role in this phase by educating aircrew members on the critical importance of mature self-discipline. This education includes emphasizing the detrimental effects of sleep loss, excessive indulgence in food, tobacco, and alcohol, the necessity of a regular physical conditioning program, and the proactive management of personal problems before they escalate into distracting "mountains of preoccupation." Stanbridge aptly suggests that a flier should approach their occupation with the same rigorous preparation as an Olympic athlete preparing for a major competition.

While most military commands implement strict policies regarding minimum crew rest periods between missions, subtle disruptions to diurnal rhythms can significantly undermine their effectiveness. These disruptions, often associated with nightly drops in core body temperature and adrenal hormone secretion, can impair the quality of rest. Individuals who regularly work night shifts typically develop a reversed circadian cycle, achieving peak performance at night and experiencing their lowest body temperature during the day. However, frequent shift changes prevent the stabilization of this rhythm, ensuring that performance remains lowest during night hours. Similarly, long-distance flights across multiple time zones induce significant sleep disturbances, as a daytime arrival at a destination may coincide with the crew's physiological nighttime.

The Directorate of Flight and Missile Safety Research has identified post-alcoholic hangover as a contributory cause in accident statistics, and an unfortunate European commercial accident recently implicated aircrew intoxication. Experimental studies have consistently shown a decrement of approximately 10% in flying-related tasks following the ingestion of as little as one ounce of alcohol, highlighting the profound impact on cognitive and psychomotor function.

Pre-flight crew control is designed to ensure that aircrews arrive at their aircraft thoroughly rested and ready for flight. Acknowledging that potential stressors such as clearance delays, inadequate transit facilities, unexpected schedule changes, and anxieties related to personal equipment can exacerbate existing fatigue potential, proactive measures are crucial.

In the Strategic Air Command, dedicated attention to crew facilities has demonstrated the tangible benefits of environmental support. The installation of hammocks, ovens for hot foods, cold chests for frozen dinners, and the provision of additional crew members to permit rotation and sleep schedules have proven instrumental in maintaining effective performance during exceptionally prolonged missions.

Intensive training is also a vital component, as it enhances confidence and fosters more relaxed flying responses. Familiarity with procedures and equipment make in-flight emergencies or Ground Controlled Approaches (GCAs) at destination airfields less anxiety-inducing. Furthermore, specific simulator-based indoctrination designed to guard against skill fatigue has led to significant performance improvements. Often, fatigue simply indicates the need for brief moments of variety and change to restore optimal performance. Analogous to industrial settings where work output with scheduled breaks is superior to continuous, uninterrupted work, strategic breaks can enhance aviation performance.

Pharmacological interventions, while potentially useful, must be managed with extreme caution. Coffee, for example, should be reserved for the latter half of a flight, as many individuals experience a "post-caffeine letdown." Similarly, meals should be nutritionally balanced, avoiding over-reliance on pure carbohydrates, which can lead to an "insulin overshoot" and subsequent delayed hypoglycemia. Fundamentally, individuals in better physical condition exhibit greater tolerance to fatigue, experience less postural tiring, and recover more rapidly after missions.

Dextroamphetamine has been administered to a large number of aircrews, with no accidents directly attributable to the drug. However, side reactions have been observed in flight. One F-100 pilot, who had shown no side effects during pre-testing, experienced euphoria and a narrowed span of attention after taking his first 5 mgm dose of Dexedrine prior to a mid-Atlantic refueling. He reported being able to concentrate on only one procedure at a time, a potentially critical impairment. Other aircrew members have noted agitation and hyperactivity. Fortunately, advancements in crew rest facilities within larger aircraft have largely obviated the requirement for stimulant medications on many missions.

In summary, while fatigue is an inherent human response to sustained stress, its detrimental effects can be significantly minimized through a robust system of crew and environmental controls. The Flight Surgeon occupies a central and indispensable role in this process, contributing to the planning, evaluation, and provision of technical advice on all fatigue-preventing systems and protocols applied to human operators in the complex aerospace environment.

Environmental and Personal Factors in Fatigue

The development and manifestation of fatigue in aircrews are profoundly influenced by a complex interplay of environmental and personal factors. These elements, though distinct, often interact synergistically to either exacerbate or mitigate an individual's susceptibility to performance decrement. Understanding these factors is crucial for Flight Surgeons in designing comprehensive fatigue management strategies.

Environmental Factors: The operational environment of military aviation presents a unique array of physical and psychological stressors that directly contribute to fatigue. Elements such as **hypoxia,** even in its mildest forms, can subtly impair cognitive functions and accelerate the onset of weariness. **Temperature extremes,** whether the scorching heat of a desert airbase or the freezing cold at high altitudes, impose significant physiological strain. Prolonged exposure to **high-intensity noise and constant vibration,** particularly within the confined spaces of aircraft cockpits, leads to sensory overload and a generalized sense of exhaustion, contributing to both physical and mental fatigue. The **sustained application of G-forces,** common in combat maneuvers, not only causes physical stress and muscle strain but also demands increased physiological effort to maintain consciousness and control.

Beyond these immediate physical stressors, the operational context introduces other critical environmental considerations: * Inability to Move: Confinement to a seat for extended periods, especially in fighter aircraft, results in physical stiffness, reduced circulation, and the tiring of postural muscles. This immobility, compounded by heavy personal equipment, frequently leads to complaints like backaches among bomber crews, a problem that has been effectively addressed with inflatable lumbar pads and innovative seating designs like reclining ejection seats and alternating inflatable cushions, as evidenced in B-47 missions. * **Feeding and Elimination:** Irregular meal schedules, unpalatable in-flight food, or difficulties with in-flight waste disposal can disrupt physiological rhythms, reduce energy replenishment, and negatively impact morale, thereby aggravating fatigue. * **Sleep Environment:** The quality and duration of sleep are paramount to recovery. Suboptimal living conditions, noisy sleeping areas (e.g., proximity to flight lines), and disruptions to diurnal rhythms (e.g., changing shifts, trans-meridian flights) severely impede restorative sleep. As observed during the Berlin Airlift, moving sleeping areas away from the flight line significantly reduced aircrew fatigue. * **Logistical Support:** Deficiencies in ground and air organizational support, such as poor transit facilities, insufficient recreational opportunities, or inadequate administrative support, all contribute to a cumulative stress burden. The RCAF Tokyo Airlift study highlighted the impact of factors like food, quarters, transportation, and duty hours on crew fatigue.

Personal Factors: Beyond the external environment, intrinsic characteristics of the individual profoundly modulate their response to fatigue-inducing conditions. * **Physical Conditioning:** An individual's baseline physical fitness is a significant determinant of fatigue tolerance. Those in better physical condition generally tolerate fatigue more effectively, experience less postural tiring, and recover more quickly after missions. Conversely, poor muscle tone exacerbates the discomforts of prolonged sitting and contributes to overall weariness. * **Nutritional Status:** While flying typically involves minimal physical exertion that would lead to rapid glucose depletion, a state of low blood sugar resulting from dietary neglect can significantly contribute to mental irritability and impaired performance. During long flights, inadequate replenishment of energy reserves invariably exacerbates fatigue. The type of meals consumed also matters; balanced meals are preferable to pure carbohydrates, which can lead to delayed hypoglycemia through an "insulin overshoot." * **Sleep Deprivation:** This is arguably the most critical personal factor. Both pre-flight and during excessively long missions, insufficient or fragmented sleep directly undermines an individual's capacity to sustain performance. The Directorate of Flight and Missile Safety Research emphasizes "time out of bed" as a more sensitive indicator of fatigue than simply duty time, noting tragic accidents linked to pilots flying at night after a full day's work. * **Hypoxia and Carbon Monoxide Exposure:** Even mild, chronic exposure to carbon monoxide, particularly prevalent in heavy smokers at altitude (leading to 8% CO-Hb saturation), can exert effects similar to hypoxia, further reducing physiological reserve and accelerating fatigue onset, especially during flights at moderate cabin altitudes. * **Motivation and Personality:** An individual's intrinsic motivation, resilience, and underlying personality structure play a crucial role in their ability to tolerate stress and resist fatigue. A strong sense of purpose and a healthy emotional constitution can significantly enhance coping mechanisms. Conversely, individuals with passive behavior patterns or underlying emotional disturbances are predisposed to fatigue-related dysfunction, particularly in the absence of unusual environmental stresses. * **Experience and Training:** Experienced pilots often develop compensatory "reflexes," such as muscle tensing, to combat G-forces, and they learn to anticipate and recognize their G limits. This acquired experience significantly enhances their resilience to the physical stressors of flight. Similarly, intensive training improves confidence and facilitates more relaxed flying responses, making in-flight emergencies or challenging approaches less worrying. Training also instills a greater awareness of the signs of skill fatigue, leading to significant performance improvement through proactive self-management. * **Alcohol and Drugs:** The consumption of alcohol, even in moderate amounts, has been shown to cause a measurable decrement in flying-type tasks and can contribute to accidents. Similarly, unauthorized use of drugs, including some over-the-counter medications, can impair judgment, depth perception, or alertness, thereby reducing fatigue tolerance. While analeptics like dextroamphetamine can postpone the need for sleep, their use must be strictly controlled due to potential side effects like euphoria, narrowed attention, agitation, and hyperactivity. They are not a substitute for adequate rest.

The nuanced understanding of how these diverse environmental and personal factors interact is essential for Flight Surgeons. By recognizing the unique vulnerabilities and adaptive capacities of each aircrew member within their specific operational context, Flight Surgeons can tailor interventions that effectively mitigate fatigue, thereby safeguarding both individual well-being and mission success.

Future Outlook: Fatigue in Space Operations

As humanity ventures further into the cosmos, the challenges of fatigue are poised to intensify significantly, making its management a paramount concern for future space operations. The unique conditions of spaceflight, characterized by prolonged mission durations, unprecedented environmental stressors, and the inherent isolation of the deep space environment, will introduce novel dimensions to the problem of human fatigue.

Future space missions are anticipated to be far more protracted than typical terrestrial aerospace operations. Journeys to distant celestial bodies will span months, if not years, subjecting crews to sustained periods of operation and confinement. This extended duration alone is a formidable fatigue accelerant. Furthermore, these missions will feature extended periods of unpowered parabolic flight en route, where external stimuli and workload may be minimal. While this might initially appear to offer respite, the associated monotony and sensory deprivation are themselves potent predispositions to cognitive and psychological fatigue. The absence of varied sensory input, coupled with repetitive tasks, can induce a state of profound mental weariness, challenging the crew's ability to maintain vigilance and focus.

The inherent isolation of space, combined with the criticality of mission success, will amplify feelings of responsibility and, potentially, anxiety. The continuous need for heightened alertness and meticulous attention to complex systems, even during quiescent phases, will exert a relentless cognitive load. In such an environment, the psychological dynamics of fatigue, including boredom, frustration, and the subconscious undercurrent of danger, will be intensified. The "conflict situation" described by Bartley and Chute, wherein an individual develops an aversion to their task due to feelings of unwillingness or inadequacy, could become particularly relevant in deep-space environments where escape or immediate relief from stress is impossible.

Maintaining an optimal state of alertness and performance over such extended periods, especially with minimal input workload, presents a prodigious task. Astronauts, like their aircrew counterparts, will need to develop robust defenses against skill fatigue through rigorous selection and extensive training. However, the unique stressors of microgravity, altered circadian rhythms due to lack of a natural day-night cycle, and potential sleep disturbances will necessitate entirely new strategies for fatigue mitigation. Sleep deprivation, already a critical factor in terrestrial aerospace fatigue, will become even more complex in an environment devoid of natural cues for rest.

The Flight Surgeon's role in this future paradigm will evolve to address these advanced challenges. Their expertise will be crucial in planning, evaluating, and advising on innovative fatigue-preventing systems tailored to the space environment. This includes not only optimizing habitat design and workload schedules but also exploring novel countermeasures for monotony, psychological support systems, and potentially advanced physiological monitoring to detect nascent signs of fatigue before they compromise mission safety.

Ultimately, while fatigue remains a fundamental human response to sustained stress, the successful execution of future space operations will depend on minimizing its impact through meticulously designed crew and environmental controls. The Flight Surgeon, leveraging their understanding of human physiology, psychology, and operational demands, will be an indispensable figure in ensuring the sustained effectiveness and well-being of humanity's pioneering space explorers.

References

Bartley, S. H., & Chute, E. (2018). "Human Fatigue and Impairment". McGraw-Hill. Domanski, T. J. (2020). "Stress Response in Jet Operations" (USAF School of Aviation

Medicine Report 57-16).

Floyd, W., & Welford, A. T. (Eds.). (2021). "Symposium on Fatigue".

Fraser, D. C. (2017). Aircrew Fatigue Study. "Flying Personnel Research Committee Reports of Great Britain", #984.

Frederik, W. S. (2019). Physiological Dimensions of Human Fatigue. "AMA Archives of Industrial Health", 20, 297-302.

Hauty, G. T., Payne, R. B., & Bower, R. O. (2016). Dextroamphetamine and Oxygen Deprivation on Work Performance. "SAM Report 56-125".

Langdon, D. E. (2015). Aviation Fatigue. "Aerospace Safety", 16(11).

McFarland, R. A. (2019). "Human Factors in Air Transportation". McGraw-Hill.

McGrath, S. D. (2018). Aircrew Fatigue During the RCAF Tokyo Airlift. "Journal of Aviation Medicine".

Ruch, T. J., & Fulton, J. F. (2022). "Medical Physiology and Biophysics".

Stanbridge, R. H. (2015). Aircrew Fatigue: Berlin Airlift Insights. "The Lancet", 261(6671), 1-3.

Strughold, H. (2016). Circadian Rhythms Post Long-Distance Flights. "International Record of Medicine and General Practice Clinics", 168, 576-579.

Whittingham, H. (2023). "Flight Time Fatigue". Flying Personnel Research Committee Reports of Great Britain, #FPRC 1037.

Chapter 10: Psychiatric Health of Flying Personnel

The dynamic and demanding environment of aviation consistently presents unique challenges to the psychological well-being of flying personnel. A comprehensive understanding of these challenges is crucial for Flight Surgeons to ensure the continued functional efficiency and safety of aircrew members. This chapter explores the historical evolution of recognizing psychiatric disorders in aviation, delves into the intricate interplay of stress, environmental factors, and individual personality, and examines specific psychological reactions encountered in both training and combat scenarios. Furthermore, it seeks to clarify the often-misunderstood concept of "fear of flying" and addresses the distinct psychiatric considerations arising in missile operations.

Historical Context and Core Concepts

The recognition of psychiatric disorders among flying personnel has a history stretching back to the early days of World War I. During this era, it became increasingly apparent that individuals could experience a significant loss of functional efficiency without any discernible physical ailment to explain the decline. Initially, efforts to understand this phenomenon predominantly emphasized physical aspects, with psychological factors receiving limited attention in terms of their potential etiological importance.

Early attempts to categorize the various factors negatively impacting an individual's flying ability led to the overarching term "flying stress." Unfortunately, this term soon became synonymous with the symptoms themselves, leading to the coining of phrases such as "aeroneurosis," "aviation neuroasthenia," and "flying fatigue." This linguistic development fostered a problematic assumption: that a distinct clinical entity had emerged, one inherently peculiar and unique to the flying situation, and that any individual exposed to this specific environment might inevitably develop this "disease."

In more recent years, the term "fear of flying" also gained currency, and similarly, it was often misidentified as a discrete clinical entity. However, from a contemporary aeromedical perspective, the fundamental psychiatric challenge confronting the Flight Surgeon is the loss of functional efficiency that is directly attributable to emotional factors. This shift in understanding underscores the transition from a purely physical or symptom-based approach to a more holistic, psychologically informed view of aircrew mental health. The core concept now is that emotional factors, rather than a unique "flying disease," are the primary drivers of impaired performance in aviation personnel.

Interplay of Stress, Environment, and Personality

To adequately comprehend the mechanisms underlying the loss of functional efficiency due to emotional factors in flying personnel, a comprehensive examination of the multifaceted influences acting upon the individual flyer is essential. These influential variables can be conveniently categorized into three primary groups: the specific stresses to which an individual is subjected, the environmental factors that modulate an individual's tolerance to these stresses, and the intrinsic personality factors, including inherent adaptability. The manifestation of a loss of functional effectiveness, stemming from a failure in emotional adaptation, is profoundly shaped by the subtle and complex interactions among these three categories. While these factors are deeply intertwined in real-world operational scenarios, they warrant individual discussion for clarity.

Stress

Stresses are elicited from a broad spectrum of situations capable of generating fear, insecurity, frustration, pain, fatigue, or any other form of tension or discomfort. It is important to note that while some situations unequivocally represent severe stress for virtually everyone, no situation possesses an absolute or universally quantifiable stress value. The concept of stress, therefore, only acquires meaningful relevance when considered in relation to the individual experiencing it. Consequently, a comprehensive assessment necessitates not only an understanding of the inherent nature and intensity of the stressor but also a nuanced appreciation of the specific personal significance it may hold for the individual involved.

Even within non-combat flying, a certain irreducible level of inherent stress is always present, stemming from the persistent threat of death and potential mutilation. The magnitude of this intrinsic stress is variable, influenced by several factors: the specific type of aircraft being operated (e.g., jet versus conventional), the presence or absence of other crew members, prevailing weather conditions, the nature of the mission profile, and the characteristics of the terrain over which the flight occurs. The financial implications, such as flying pay and the elevated costs of life insurance for aviators, tangibly reflect these inherent dangers, even though flyers themselves rarely consciously acknowledge these risks.

In combat situations, additional and often intensified stressors are, naturally, superimposed upon these baseline factors. Beyond the direct threat posed by enemy aggressive and defensive actions, further elements contribute to both the quantity and quality of stress. These include the anticipation of upcoming missions, the cumulative impact of casualty rates, the extended duration of mission tours, and the personal experience of injury or death among crew members and close friends. Historical data from World War II strikingly demonstrated a direct correlation between the number of aircraft lost and a corresponding rise in the incidence of emotional disorders among flying personnel. This historical evidence underscores how an elevated casualty rate directly resulted in appreciable anxiety across the aircrew.

Environmental Factors

Environmental factors hold a position of secondary importance only to specific personality factors in shaping an individual's capacity to tolerate stress, both in combat and non-combat flying contexts. These environmental influences encompass abstract yet profoundly impactful concepts such as morale, leadership efficacy, a collective sense of support, and the strength of group identity. The direct and powerful influence of these variables on an individual's resilience to stress is frequently underestimated. While military leaders have historically recognized their importance, the full significance of their implications for aeromedical practice has only gained comprehensive appreciation in more recent times.

Contemporary understandings of human behavior explicitly acknowledge the deep and potent effects of cultural norms, societal attitudes, and prevailing mores on individual conduct. It is recognized that an individual's social needs are nearly as compelling as their personal needs, and indeed, in most instances, these two domains are intricately interwoven. Thus, when an individual operates within a unit environment where social interactions generate minimal personal anxiety and frustration, while simultaneously offering maximal gratification and a sense of security, their inherent resistance to stress is significantly enhanced.

The process of "identifying with a group"—that is, developing a genuine sense of "belonging" and loyalty—serves to broaden an individual's sense of duty beyond their immediate self, extending it to encompass a responsibility towards the collective. Consequently, when the group's mission objectives are clear and understood, the individual becomes more willing and capable of tolerating increased personal stress in pursuit of that shared group goal. Conversely, strong cultural restrictions against the expression of destructive or hostile aggression can inadvertently impede the successful achievement of military objectives. One crucial cultural function of a military group, therefore, is to foster an atmosphere where aggression directed towards the enemy is not only encouraged but actively rewarded. When this is successfully achieved, individuals who might ordinarily have their aggressive impulses inhibited by guilt become more capable of expressing such aggression in appropriate contexts.

Furthermore, aggressive action possesses a secondary benefit: it can provide an effective outlet for the relief of tension and anxiety, thereby enhancing an individual's overall stress tolerance. Conversely, stressful situations that preclude active aggressive action on the part of the individual are prone to generate increased tension and a diminished capacity for stress tolerance. Illustrative examples of such challenging scenarios include executing a bomb run through intense flak, enduring an prolonged air raid passively, or being grounded by adverse weather conditions when the mission called for aggressive sorties. In all these circumstances, where active engagement is thwarted, resistance to stress is notably reduced.

A robust "sense of support" emerges when both individual members and the collective group perceive that their peers are contributing equitably, and that their personal efforts and sacrifices are acknowledged and valued. This sense is cultivated by numerous, often subtle, factors. The provision of adequate living quarters, access to good quality food, the availability of recreational facilities, and a variety of other "fringe benefits" all directly contribute to this critical sense of support.

Effective "leadership" plays an indispensable role in cultivating group cohesion and in clearly delineating both individual and collective roles and responsibilities. It stands as a pivotal factor in the development of group mores, shaping attitudes, defining goals, and fostering motivation. Moreover, the crucial sense of support within a unit is largely mediated through the actions and presence of its leaders.

Collectively, all these environmental factors contribute significantly to "morale," a term that, in its general sense, refers to the degree of willingness within a group to actively strive towards achieving its collective goals. It becomes clear, therefore, that these environmental factors exert a direct influence on an individual's inherent ability and willingness to tolerate stress, making them legitimate and essential considerations within the scope of aeromedical practice.

Personality Factors

The intrinsic "personality" of the individual, upon whom the myriad of aforementioned stressors and environmental influences converge, is, of course, of fundamental and paramount importance. An individual possessing a sound basic personality structure, characterized by minimal underlying psychological conflict and a low baseline of latent anxiety, will inherently exhibit a greater capacity to tolerate stress. Core attitudes towards concepts such as aggression, military regimentation, personal self-sacrifice, and an unwavering devotion to duty will collectively shape and influence their overall performance under pressure.

Other significant personality attributes also play a critical role in stress resilience. These include an individual's capacity to genuinely identify with a group, their ability to establish and maintain effective interpersonal relationships with peers, an innate sense of responsibility, and a steadfast devotion to core principles. These cumulative traits are fundamental in determining "motivation," which stands as the key factor dictating the ultimate amount of stress an individual can, or indeed will, tolerate. Complementary personal factors, such as age, nutritional status, and the pervasive impact of fatigue, also significantly influence stress tolerance. The comprehensive effects of fatigue are, however, addressed in greater detail in a separate dedicated chapter.

Furthermore, the "opportunity for aggressive action" and relevant "training" serve to mitigate anxiety. By actively facilitating and emphasizing training and education in areas such as weapon proficiency, the execution of emergency procedures, self-defense techniques, survival strategies, and escape and evasion tactics, the inherent anxiety induced by a state of passivity can be significantly offset. This proactive approach directly enhances an individual's capacity for aggressive action, which, as previously noted, can relieve tension. Education, in this context, functions to dispel the inherent doubts and anxieties associated with unknown or fantasied perils. It replaces these uncertainties with informed expectation and a robust sense of confidence, which arises from the mastery of pre-planned offensive and defensive measures.

In summary, a diverse array of factors demonstrably contributes to an individual's overall resistance to stress. These multifaceted influences must be meticulously considered by the Flight Surgeon to achieve optimal effectiveness in assisting aircrew members in adapting to challenging situations. Consequently, the professional role of the Flight Surgeon extends far beyond merely recognizing and treating individual reactions to stress. It inherently encompasses a profound understanding and a thorough analysis of the multitude of other factors operating within their designated unit, thereby ensuring a holistic approach to aircrew mental health.

Common Psychiatric Reactions in Flight Training

The psychological reaction types observed in flying personnel, particularly during the initial stages of training, do not significantly diverge from those commonly encountered in civilian populations. These reactions can be accurately described and comprehensively explained through the application of modern psychiatric terminology and established psychopathology, providing both diagnostic clarity and dynamic insight. This holds especially true for the early phases of flight training, during which the intensity of environmental stress is typically kept to a minimal level.

Despite the implementation of rigorous selection methods designed to identify and screen suitable candidates, a certain proportion of accepted individuals will, nonetheless, manifest psychological dysfunction early in their training curriculum. A variety of reaction types are commonly observed, with several standing out as particularly prevalent. These include: * **Somatization-reactions:** These manifest as physical symptoms for which no adequate medical explanation can be found. Common presentations include persistent headaches, chronic back pain, and various forms of gastrointestinal dysfunction. * **Anxiety-reactions:** Characterized by excessive worry, apprehension, nervousness, and often accompanying physical symptoms of autonomic arousal. * **Conversion-reactions:** These typically involve neurological symptoms, such as disturbances in the special senses like vision (e.g., blurred vision, tunnel vision) or hearing (e.g., sudden hearing loss, tinnitus), without any underlying organic pathology. * Psychogenic motion sickness: This is motion sickness that is primarily psychological in origin, rather than a direct physiological response to motion. *

Character and behavior disorders: These involve ingrained patterns of thinking, feeling, and behaving that deviate significantly from cultural expectations, leading to distress or impairment in functioning.

Individuals whose symptoms emerge early in their training, or in the absence of any particularly unusual or intense environmental stressors, are frequently found to possess underlying immature or passive behavioral patterns. Such personality traits often predispose them to developing emotional disturbances when confronted with new challenges. It is plausible that some individuals might embark on a flying career with subconscious motivations, such as a desire to prove their masculinity or to deny a deep-seated, perhaps unacknowledged, need for dependence. In the majority of such cases, it becomes necessary to recommend their removal from flight training, given the high likelihood that these symptoms would recur and intensify under the increased stresses anticipated in future operational contexts.

Psychological Responses to Combat Stress

Under the extreme pressures of flying hazardous combat missions, the presence of a certain degree of apprehension and fear is not only expected but almost universally observed among aircrew. Furthermore, many individuals will experience various somatic manifestations that are direct concomitants of this intense apprehension. The physiological effects of anxiety, primarily mediated through the autonomic nervous system, can include a range of symptoms: tachycardia (rapid heart rate), hyperventilation, sensations of nausea or "queasiness," diarrhea, increased urinary frequency, tremulousness (shaking), and an exaggerated "startle" response. Individuals affected by these responses may also report feeling restless and irritable, and may suffer from insomnia and anorexia (loss of appetite). These symptoms are so common and pervasive in the combat environment that they are, in fact, often considered "normal" reactions to the extraordinary demands of warfare.

The majority of individuals, sustained by their strong intrinsic motivation and the robust supportive dynamics within their group's environment, are able to tolerate these challenging symptoms. They continue to operate effectively despite their internal discomfort. However, there is a distinct subset of individuals—an occasional person—who, driven by an innate reluctance to persist in the hazardous situation, will actively seek removal. This desire for removal is frequently rationalized and justified on the basis of their physical symptoms. These individuals often have completed only a limited number of missions and may not have been directly exposed to any overtly unique or unusually stressful experiences when their symptoms begin to manifest significantly. Such an individual is highly likely to approach the Flight Surgeon with a pre-conceived conviction that grounding or hospitalization is a necessary and unavoidable course of action.

For some of these individuals, continuation in their flying duties may still be possible. This often occurs if the underlying cause and the mechanisms of their symptoms are thoroughly explained to them. Crucially, the Flight Surgeon's manner must convey an unwavering underlying attitude that the individual, despite these entirely normal manifestations of apprehension, possesses the capacity to continue. However, if these initial, more superficial interventions prove insufficient, and the individual's removal from flying duties becomes unavoidable, this disposition must be handled on an administrative basis. To pursue any other course, particularly a medical one, would invariably risk undermining the moral resolve and continued functional capacity of other aircrew members who are actively struggling and performing effectively despite their own fears and discomfort.

A separate group of aircrew members develops symptoms only later in their tour of duty, after having successfully flown a substantial number of stressful missions. These individuals recognize their growing reluctance to continue, yet crucially, due to their strong underlying motivation and deep identification with their group, they express a genuine desire to complete their assigned tours. Consequently, they often respond favorably to the supportive and explanatory measures previously mentioned. If, however, chronic anxiety and fatigue have progressed to the point of causing persistent insomnia, significant weight loss, or other tangible physical manifestations, the individual may benefit from more direct supportive interventions. These could include short-term administration of sedatives for a few days, or a brief "rest and recuperation" leave designed to provide a period of respite and recovery.

Crucially, these sensitive dispositional decisions are most effectively made by the unit Flight Surgeon. If a pilot is evacuated to a distant hospital for a decision, the vital ties connecting the individual to their operational unit are severed, resulting in the loss of a powerful motivating force. Furthermore, a medical officer who is not integrated within the unit lacks the intimate identification with the group and its specific operational context that the local Flight Surgeon possesses. Such a distant clinician's understandable empathy for an unhappy patient might inadvertently lead to the loss, via medical channels, of a pilot who, with appropriate unit-level intervention, could have successfully continued their duties. The unit Flight Surgeon, by virtue of being an integral member of the operational group, is uniquely positioned to recognize that an overly permissive or "too-easy" release of an individual pilot from their responsibilities, through medical means, could ultimately erode the cohesion and effectiveness of the group as a whole.

Conversely, it must also be acknowledged that an unduly "tough" or unyielding policy regarding such cases can similarly prove detrimental to group morale. When an individual has been subjected to prolonged and severe hazards, and ultimately becomes incapacitated despite possessing fundamentally good motivation and a strong desire to continue, then a medical disposition is genuinely indicated. In these specific circumstances, other flyers can readily empathize with the disabled pilot's plight. Therefore, a policy perceived as excessively rigid or unsympathetic could, in fact, paradoxically contribute to a deterioration in overall group morale. The decision concerning a medical versus an administrative disposition is rarely straightforward; it demands a high degree of mature professional judgment on the part of the physician, balancing individual welfare with unit effectiveness.

"Fear of Flying": Clarifying Medical vs. Administrative Perspectives

The term "Fear of Flying" has historically been employed in both medical and administrative contexts, frequently without a clear distinction between these two very different interpretations. This ambiguity has unfortunately led to significant confusion within this critical area of aeromedical practice.

Complex questions invariably emerge when a flyer develops incapacitating emotional symptoms that overtly threaten flying safety and appear uniquely linked to the stresses of aviation. A central dilemma for the Flight Surgeon involves discerning the root cause of these disabling symptoms: Are they indicative of a "genuine" underlying illness, genuinely impeding the effectiveness of a conscientious and responsible individual who is actively striving to overcome their difficulties? Or, conversely, do these symptoms reflect the behavior of a poorly motivated, less responsible individual who finds the prospect of "secondary gain" (i.e., removal from a stressful situation) too compelling to resist suppressing their uncomfortable symptoms? To grant medical grounding for symptoms primarily driven by poor motivation would be profoundly unfair to the rest of the aircrew, many of whom may be operating with equal or even greater levels of anxiety and discomfort. This could severely impact morale and a sense of equity within the unit. Conversely, to "punish" a sincere and dedicated flyer who has genuinely reached their psychological limits, having endured more than they can reasonably sustain, would also be profoundly unfair to both that individual and the broader group. In either scenario, the morale and motivation of the entire group are susceptible to adverse effects.

As previously established, an individual's tolerance to stress is not static; it is a variable influenced by a multitude of factors, among which group pressures and attitudes are highly significant. Voluntary removal from flying status can, in some military cultures, lead to social disapproval of the individual. Being actively engaged in flying, particularly when it entails considerable personal risk and sacrifice, carries substantial prestige. However, such cultural values must not be permitted to unduly influence medical judgment or further complicate an already challenging situation. It is precisely in this intricate domain that the skill and clinical acumen of the Flight Surgeon are most rigorously tested. The evaluation of many such cases is inherently difficult, as it demands considerable subjective judgment to determine what constitutes "enough" sustained stress and what defines genuinely satisfactory motivation.

Current Air Force regulations specifically define "fear of flying" as encompassing all emotional reactions that are exclusively related to flying activities and are absent when not under flying stress (USAF, 2023). These regulations make provisions for psychiatric treatment, and the individual may be reinstated to flying status if and when they are deemed physically and psychologically qualified. However, it is herein posited that the term "fear of flying" should be judiciously reserved to characterize individuals primarily exhibiting faulty motivation. It should not be indiscriminately applied to those experiencing true psychiatric illnesses, irrespective of the relationship of their symptoms to flying. While it is certainly true that conscious "fears" of flying may be present as a symptom within genuine psychiatric conditions, such as phobic disorders or anxiety reactions, this fear is merely a symptom and not the underlying disease itself. To label such reactions solely as "fear of flying" risks confusing them with cases that are more appropriately managed through administrative channels, given their motivational rather than pathological origin.

It must be emphasized that the mere presence of emotional symptoms does not automatically equate to psychiatric disease; these symptoms can, and frequently do, reflect underlying issues of faulty motivation. Therefore, the fair and judicious handling of each individual case necessitates a meticulous appraisal of the confluence of stresses, environmental influences, and motivational or personality factors that have contributed to the formation of symptoms. Disposition decisions must not only prioritize what is deemed best for the individual's well-being but also, crucially, consider which course of action will most positively impact the successful accomplishment of the group's mission. In practical application, the needs of the individual and the objectives of the group are typically quite compatible. This compatibility stems, in part, from the understanding that allowing a poorly motivated person an unearned exit from flying duties through medical channels could inadvertently saddle them with chronic feelings of failure, guilt, and anxiety. Conversely, a firm yet supportive approach to management often leads to an early and successful return to flying duties, an outcome that is optimal for both the individual's career progression and the sustained effectiveness of the operational group.

Psychiatric Challenges in Missile Operations

The psychiatric landscape in missile operations presents a distinct set of challenges, wherein the relative influence of direct stress and environmental factors upon the individual differs notably from that encountered in traditional flying situations. While certain stresses are inherent—such as the hazards associated with explosions, exposure to noxious fumes, and the risk of accidents—these are generally perceived as less acute and, perhaps paradoxically, less anxiety-provoking than the immediate perils of combat flying. The chronic, high-intensity combat tensions typical of aerial warfare are largely absent in missile command and control environments.

Conversely, environmental factors assume a considerably greater significance in missile operations. These include: * **Remote Locations:** Missile sites are often situated in isolated, geographically distant areas, which can contribute to feelings of detachment and restrict access to typical social and recreational outlets. * **Small Stations with Few Personnel:** The limited number of personnel at these sites can lead to a sense of social isolation and intense interdependencies, potentially amplifying interpersonal conflicts or the impact of individual dysfunctions. * **Limited Recreational Facilities:** Restricted access to leisure and recreational activities can exacerbate boredom, reduce opportunities for stress relief, and contribute to overall dissatisfaction. * **Insufficient and Potentially Substandard Housing:** Inadequate or uncomfortable living conditions further compound the psychological toll, undermining morale and a sense of support.

All these variables are crucial in shaping the mental health of missile crews. The overarching strategy of deterrence relies fundamentally on the formidable destructive potential of our missile systems. The very premise dictates that should these weapons ever need to be launched, the strategy of deterrence would inherently have failed. Consequently, the ability to operate these complex weapon systems effectively, particularly under conditions of extremely short warning times, is absolutely essential for the success of this strategic posture. This requirement for constant vigilance, therefore, demands an individual who is not merely alert but perpetually "keyed-up."

Maintaining such an elevated state of mental readiness is inherently challenging, even in the context of a "hot war" where the imminent use of the weapon system is a tangible and immediate threat. It becomes a truly prodigious task to counteract the insidious onset of complacency and to sustain an alert attitude within a "cold war" paradigm, where the successful functioning of the weapon system effectively means it never has to be operated. These specific contributing factors—namely, the remote geographical location, the psychological demands of a "cold war" environment, and the inherent lack of operational opportunity to physically engage the system—collectively render individuals more susceptible to impaired efficiency stemming from emotional symptoms.

Furthermore, empirical data has underscored the critical human element in these sophisticated systems: it has been demonstrated that approximately 25% to 40% of all missile failures are directly attributable to human error. A significant proportion of these errors, in turn, can be traced back to impaired efficiency resulting from underlying emotional tensions. Phenomena such as momentary lapses of attention, simple operational mistakes, and what might be perceived as slipshod or careless work can often be directly linked to the cumulative effects of psychological pressures.

Given these profound implications, counteractive measures in missile operations must be as robust, or even more so, than those applied in combat flying situations. The aim is to rigorously guard against significant decrements in operational effectiveness. The Flight Surgeon's efforts in this unique context should be strategically directed towards two key areas: first, the initial elimination, through rigorous screening, of individuals whose presence is deemed detrimental to the group or who may pose a threat to security; and second, the continual screening and monitoring of personnel to ensure ongoing suitability. The Flight Surgeon must also actively collaborate with command leadership to identify and address morale-reducing factors and any instances of faulty leadership that might undermine unit cohesion. Furthermore, they are uniquely positioned to advocate for and demonstrate the critical need for improved living conditions, enhanced recreational facilities, and better transportation options for personnel stationed at these remote sites.

As in combat situations, the Flight Surgeon must exercise considerable medical judgment in determining who should be transferred from a remote missile site. Both excessively lenient and overly strict approaches can have a detrimental impact on the unit's overall effectiveness. Therefore, the local Flight Surgeon, mirroring their critical role in historical contexts, remains in the optimal position to evaluate each individual case with nuanced understanding and to arrange for an appropriate disposition. This approach not only ensures fairness to both the individual and the group but, critically, sustains group effectiveness at the highest possible operational level.

References

USAF. (2021). "Medical Examination (AFM 160-1)". U.S. Air Force.

Armstrong, H. G. (2018). Neuropsychiatry in Aviation. In "Aerospace Medicine" (Chapter 23). Williams and Wilkins Co.

Mebane, J. C. (2017). "Neuropsychiatry for Flight Surgeons" (School of Aviation Medicine, USAF).

USAF. (2023). "Fear of Flying (AFR 36-70)". U.S. Air Force.

Bond, D. D. (2015). "The Love and Fear of Flying". International University Press, Inc. Grinker, R. R., & Spiegel, J. P. (2019). "Men Under Stress". Blakiston.

Glass, A. J. (2016). Combat Zone Psychotherapy. "American Journal of Psychiatry", 110, 725-731.

Stafford, C. P. (2020). Morale and Flying Experience: Wartime Study Findings. "Journal of Mental Science", 95, 10-50.

Chapter 11: Aircraft Toxicology: ManagingHazardous Gases and Vapors

Flight Surgeons and other U.S. Air Force (USAF) medical officers must maintain a keen awareness of toxic gases and vapors that may, under specific circumstances, infiltrate crew and passenger compartments of aircraft. Although stringent care is exercised in aircraft manufacturing and design, unusual operational conditions or material degradation can lead to the permeation of these hazardous substances into occupied areas. The resulting physiological effects are dangerous, stemming from a combination of the inherent toxicity of the agents, their concentration, and the duration of exposure.

Exposure to toxic chemicals in flight is typically brief; however, in longer missions, particularly those involving heavy bombers, exposure duration can be extended. Crucially, unlike ground personnel who may evacuate contaminated areas, aircrew and passengers are often confined to the aircraft until a safe landing can be executed. Consequently, it is paramount that flying personnel possess a comprehensive understanding of the toxic chemicals they might encounter. They must cultivate an acute awareness of the potential presence of these toxic vapors and be proficient in implementing appropriate emergency measures when necessary.

Contamination of the aircraft atmosphere can arise from various sources, including exhaust gases, hydraulic fluid mist, fuel vapors, coolant fluid vapors, oil vapors, anti-icing fluid vapors, fire extinguishing fluids, cargo emissions, and thermal decomposition products from electrical insulation. Each of these sources presents a unique toxicological profile that necessitates specific preventative and responsive strategies.

Exhaust Gases: Piston vs. Jet Engines

The composition of exhaust gases is highly variable, primarily influenced by the engine type and the fuel-air ratio at which it operates. Understanding these differences is crucial for assessing potential hazards.

Piston Engines

For piston engines, the approximate percentage compositions of exhaust gases by weight are illustrative of the operational conditions. For example, during take-off (with a fuel-air ratio of 0.095), carbon monoxide (CO) typically comprises 8.75%, carbon dioxide (CO_2) 10.14%, methane (CH_4) 0.31%, ethylene (C_2H_4) 0.37%, nitrogen (N_2) 70.43%, oxygen (O_2) 0.84%, and water (H_2O) 8.80%. In cruising flight (with a fuel-air ratio of 0.075), CO drops to 3.03%, while CO_2 increases to 15.11%, with corresponding minor changes in other components.

Carbon monoxide, methane, and hydrogen are products of incomplete fuel combustion. As the fuel-air ratio decreases, leading to more complete combustion, the percentage of carbon dioxide in the exhaust increases, while carbon monoxide levels decline. Conversely, a richer fuel mixture results in an increased concentration of carbon monoxide in the exhaust gases.

Aircraft designs inherently influence the frequency and severity of exhaust gas contamination. Single-engine piston aircraft, with engines situated directly forward of the fuselage, are generally more susceptible to contamination than multi-engine aircraft with laterally positioned engines. Furthermore, liquid-cooled single-engine types are less prone to exhaust gas contamination compared to air-cooled radial engine airplanes.

Jet Engines

Jet engine exhaust gases predominantly consist of over 95% air, with the remainder being almost entirely carbon dioxide. The probability of encountering toxic levels of carbon monoxide from jet engine exhaust is consequently remote. However, jet fuels are permitted to contain significantly higher sulfur levels than gasoline. This can lead to the presence of irritating concentrations of sulfur dioxide and aldehydes in the exhaust gases.

While there are limits to the mercaptan sulfur content in jet engine fuel, the vapors of liquid jet fuel may possess an unpleasant odor due to less extensive "sweetening" (a chemical process that removes mercaptans) compared to gasoline.

All new aircraft models are subject to rigorous specifications regarding freedom from carbon monoxide contamination before being accepted by the Air Force. Nevertheless, wear and tear, or structural modifications introduced during service, can compromise an aircraft's original freedom from contamination.

Periodic tests are therefore essential to detect such contamination and verify the adequacy of maintenance services.

Carbon Monoxide: Toxicity, Symptoms, and Management

Carbon monoxide (CO) is a critical toxicological concern in aviation due to its widespread presence in combustion products and its potent physiological effects.

Detection

The presence of carbon monoxide should be suspected whenever fumes indicative of heater or exhaust sources are noted. As CO itself is odorless, its detection relies on specialized equipment. Readily available options include the Mines Safety Appliance carbon monoxide indicator or colorimetric methods developed by the National Bureau of Standards (e.g., Detector, carbon monoxide, Type B-l, 6685490-2022).

Allowable Concentration

The maximum allowable concentration of carbon monoxide in Air Force cockpits is established at 0.005%, equivalent to 50 parts per million (ppm).

Monitoring

Tests of cabin air provide only a snapshot of conditions at a specific moment. These conditions fluctuate based on engine running time, fuel-air ratio, ventilation, and the sampling location within the cabin. From a practical standpoint, the pilot's blood carbon monoxide content offers a more comprehensive assessment, as it reflects the cumulative effect of gas exposure.

Blood analysis for carbon monoxide can be performed using methods such as the Van Slyke apparatus or the microtechnique by Scholander and Roughton. However, these techniques are demanding and require experienced technicians for reliable results; standard hospital clinical chemistry laboratories are often not equipped to provide consistently accurate blood gas analyses. Importantly, a poorly executed determination can lead to dangerous misinterpretations. Many "quantitative" tests performed in average laboratories cannot distinguish between hazardous blood CO levels and the baseline levels commonly found in smokers.

When evaluating blood concentrations, it is essential to consider "normal" values, especially in smokers, whose control values for carbon monoxide saturation can be as high as 8%. The Environmental Health Laboratories at Kelly AFB, Texas, or McClellan AFB, California, offer specialized services for blood carboxyhemoglobin determinations (AFR 160-15).

Causes in Cockpit

Carbon monoxide can infiltrate the cockpit through various pathways. Failures in the exhaust system, such as cracks in exhaust stacks from excessive vibration or worn packings around collector rings, have been implicated in several contamination incidents. In aircraft equipped with exhaust heaters, CO contamination can result from wear of the intensifier tube assembly or defects caused by enemy fire; consequently, pilots are advised against using exhaust heaters in combat scenarios. While incomplete oxidation of explosive mixtures can produce CO in turrets and near gun positions, the firing of guns and cannons has not been a significant source of the gas in Air Force aircraft.

Pharmacology

Carbon monoxide is a colorless, odorless gas, slightly lighter than air. Its lack of odor necessitates vigilance for exhaust gas smells as a warning sign. CO is absorbed exclusively through the lungs, with the rate of uptake influenced by factors such as respiration rate and depth, CO concentration in the air, exposure duration, blood volume, hemoglobin concentration, and the existing level of blood CO saturation.

For an individual with a normal hemoglobin concentration of 16 grams per 100 cc, 20 volumes percent of carbon monoxide represents complete saturation of the blood. At blood saturations not exceeding 35%, the rate of uptake can be approximated by the formula: A % COHb = (total ventilation * CO concentration of air) / Blood volume. Assuming a blood volume of 6,000 cc and a respiratory rate of 20 liters per minute (moderate exercise), if the permissible blood carbon monoxide is limited to 10% saturation, then 20 parts CO per 10,000 can be breathed for 5 minutes; 15 parts CO per 10,000 for 6.5 minutes; and 10 parts CO per 10,000 for 10 minutes.

For an individual at rest (ventilation 6 liters/minute, pulse 70), the increase in

%COHb is approximately (%CO * t) / 0.3, where %CO is the concentration in inspired air and t is minutes of exposure. This divisor changes for varying activity levels: 0.2 for light activity (9.5 liters/minute, pulse 80), 0.12 for light work (18 liters/minute, pulse 110), and 0.085 for heavy work (30 liters/minute, pulse 135).

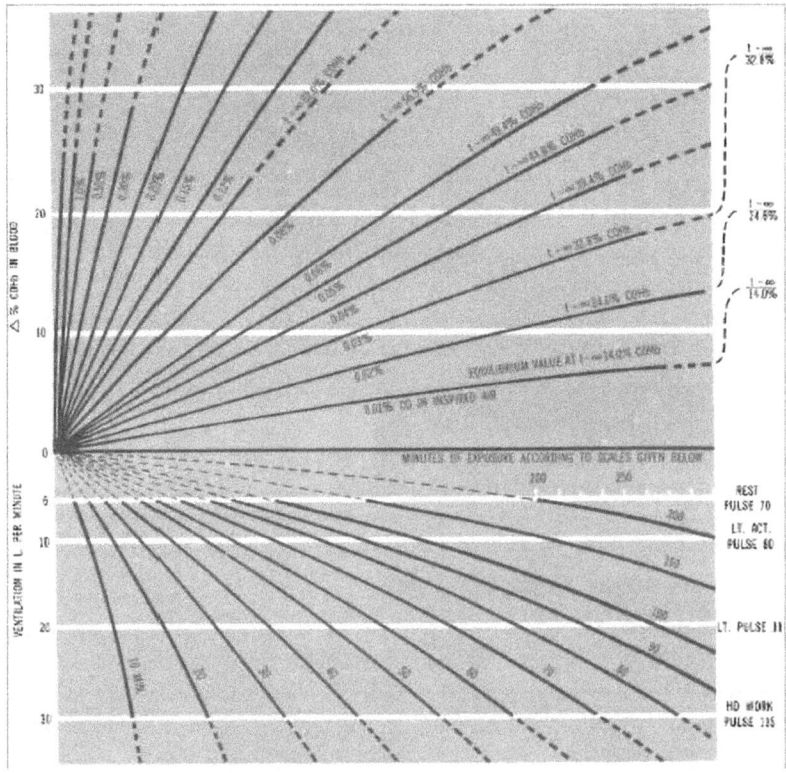

Initially, about 50% of inspired carbon monoxide is taken up by the blood at rest or during light activity. Of the gas that reaches the alveoli, a much larger proportion, approximately 90%, is retained. The affinity of human hemoglobin for carbon monoxide is 210 to 300 times greater

than its affinity for oxygen. The formation of carboxy-hemoglobin is promoted by a reduction in oxygen concentration in the air, or by increases in temperature, humidity, or physical activity.

Strictly speaking, CO is not a poison but a tissue asphyxiant, acting through a dual mechanism. First, it combines with hemoglobin, partially excluding oxygen uptake by the blood. Second, it causes a leftward shift of the oxygen dissociation curve of the remaining hemoglobin, making the curve less S-shaped and more hyperbolic (Haldane effect). This means that hemoglobin partially saturated with CO holds onto its oxygen more tenaciously, reducing oxygen liberation to tissues. Both mechanisms contribute to cellular hypoxia.

Symptoms

The physiological effects of CO poisoning primarily target structures most sensitive to anoxia, notably the central nervous system and the myocardium. The leading symptoms, in order of frequency, include headache, weakness, vertigo, nervousness, dyspnea, paresthesias, muscular twitchings, emotional disturbances, nausea, drowsiness, unsteady gait, neuromuscular and joint pains, tremors, muscular cramps, coughing, sweating, vomiting, insomnia, anorexia, precordial distress, vasomotor instability, perversion of taste and smell, impairment of speech and hearing, hoarseness, and yawning. Ophthalmological manifestations, particularly after prolonged or repeated exposure, can include contraction of visual fields, amblyopia, anisocoria, retinal edema, diplopia, and neuroretinitis, which are highly significant for flying personnel. In acute poisoning, consciousness is rapidly lost, leading to fewer reported symptoms. The greatest variability in symptoms is observed in cases involving less severe, protracted, and repeated exposures.

Blood concentrations of CO up to 10% saturation typically cause no symptoms under normal conditions (sea level, moderate physical activity, normal hemoglobin). As saturation increases, symptoms emerge in a predictable sequence: 10% saturation causes shortness of breath on vigorous muscular exertion; 20% results in shortness of breath on moderate exertion and slight headache; 30% presents with decided headache, fatigability, irritability, and impaired judgment; 40% to 50% leads to headache, confusion, collapse, and fainting; 60% to 70% results in unconsciousness, respiratory failure, and potential death if exposure is prolonged; and 80% or more saturation is rapidly fatal. The approximate times for symptom appearance with varying CO concentrations are well-documented, indicating rapid onset of severe effects at higher concentrations.

% CO in Air	Effects
0.02	Possibly headache, mild frontal in 2 to 3 hours.
0.04	Headache, frontal, and nausea after 1 to 2 hours; occipital after 2½ to 3½ hours.
0.08	Headache, dizziness and nausea in ¾ hour; collapse and possibly unconsciousness in 2 hours.
0.16	Headache, dizziness and nausea in 20 minutes; collapse, unconsciousness, possibly death in 2 hours.
0.32	Headache and dizziness in 5 to 10 minutes; unconsciousness and danger of death in 30 minutes.
0.64	Headache and dizziness in 1 to 2 minutes; unconsciousness and danger of death in 10 to 15 minutes.
1.28	Immediate effect; unconsciousness and danger of death in 1 to 3 minutes.

(Reprinted from Industrial Toxicology by Hamilton and Hardy, Second Edition, by permission of Paul B. Hoeber, Inc., Medical Book Department of Harper & Brothers, New York. Copyright 1949.)

The hazard of carbon monoxide significantly escalates at altitudes above sea level. Even mild hypoxia induced by altitude, when combined with small amounts of CO (which might be harmless alone), can synergistically cause serious impairment of efficiency due to additive hypoxic effects. For instance, if a minimum blood O2 saturation of 85% is required for maintaining flying efficiency, a CO concentration of 0.01% (100 ppm), relatively safe at ground level, reduces blood oxygenation by 10.5%. This reduction, superimposed on the naturally decreased O2 saturation at 10,000 feet, creates a dangerous hypoxic state. This situation is particularly critical for heavy smokers, who may start with a baseline of 8% carbon monoxide saturation.

Above 10,000 feet, the risks associated with carbon monoxide diminish with increasing altitude when a demand oxygen system is used. This is because, as the percentage of oxygen supplied by the demand system increases with altitude, less atmospheric air (and thus less CO) is inhaled. Consequently, while alveolar partial pressure of O2 (pO2) is maintained constant due to its proportionately larger volume in the inspired mixture, alveolar partial pressure of CO (pCO) declines due to both its decreasing percentage and the increasing altitude. Above 30,000 feet, where the demand system delivers 100% oxygen, the action of carbon monoxide is completely mitigated.

Elimination

Once absorbed, some carbon monoxide is chemically converted within the body, while the remainder is exhaled unchanged. The rate of elimination is primarily governed by respiratory volume and the percentage of oxygen in the inspired air. After absorbing moderate amounts of CO, an individual breathing pure air at sea level will clear about half of the gas from their blood within one hour, with virtually complete elimination occurring within 8 hours. Increased oxygen concentrations significantly accelerate this process; breathing 100% oxygen can reduce elimination time to an hour or less. Experimental studies, notably on dogs and guinea pigs, have demonstrated that hyperbaric oxygen therapy (up to 3 atmospheres) can rapidly restore consciousness in animals asphyxiated by carbon monoxide.

Prophylaxis

Flying personnel who suspect the presence of carbon monoxide in the aircraft, either due to the odor of exhaust gases or the onset of symptoms such as headache, nausea, dizziness, or dimming vision, should immediately take preventative action. This includes turning off any exhaust heaters in use and donning oxygen masks, ensuring the Auto-mix regulator is set to "off" or "100% oxygen." This action effectively excludes cockpit air, providing protection from carbon monoxide.

Treatment

For individuals experiencing carbon monoxide asphyxia, particularly if breathing is weak or has ceased, definitive medical treatment involves artificial respiration, the administration of 100% oxygen, and the application of warmth to the patient while maintaining rest.

Oxides of Nitrogen: Hazards and Treatment

Oxides of nitrogen represent another significant toxicological concern, particularly in certain aviation contexts.

Exposure Sources and Composition

Personnel working in and around aircraft equipped with jet assist take-off (JATO) units that utilize red or white fuming nitric acid as an oxidizing agent for rocket fuel may be exposed to oxides of nitrogen. The specific oxides generated from nitric acid or its combustion with fuel include nitrous oxide (N2O), nitrogen oxide (NO), nitrogen dioxide (NO2), and nitric tetroxide (N2O4). These oxides are also produced when fuming nitric acid reacts with organic materials, or when it is exposed to air or diluted with water.

Allowable Concentration

The maximum allowable safe concentration for an 8-hour daily exposure to nitrous fumes is approximately 5 ppm. Dangerous levels for short-term exposure (e.g., 0.5 to 1 hour) are around 100 ppm.

Relative Importance

From a toxicological standpoint, nitrogen dioxide (NO_2) is the most critical among the oxides of nitrogen. Nitrous oxide (N_2O) and nitrogen oxide (NO) are relatively less significant in typical combustion scenarios, as their concentrations are usually too low to produce characteristic toxic effects. Nitrous oxide, for instance, exhibits a narcotic effect on the central nervous system only when inhaled in high, undiluted concentrations. Nitrogen oxide can induce anoxia by forming methemoglobin and by depressing the respiratory center, and while its depressant effect in mice is more rapid and pronounced than N_2O, no human poisonings from NO have been reported. In contrast, numerous cases of poisoning from nitrogen dioxide have been documented.

Pharmacology

Nitrogen oxides offer little sensory warning; an individual may breathe an atmosphere containing lethal concentrations of nitrous fumes without experiencing serious discomfort, only to succumb hours later. When nitrogen dioxide is inhaled at body temperature, it rapidly converts to a mixture of approximately 30% NO_2 and 70% N_2O_4. The latter then reacts with water to form nitric acid (HNO_3) and nitrous acid (HNO_2). These acids are responsible for the irritation of the mucous membranes of the eyes and upper respiratory tract observed after exposure to nitrogen oxide fumes.

Symptoms

Acute nonfatal cases of nitrogen oxide exposure typically manifest with headache, dizziness, cough, palpitations, and sometimes cyanosis, restlessness, and insomnia. At certain concentrations, individuals may report feeling "doped." This narcotic or depressive action can suppress the cough reflex, leaving the patient unaware of the imminent danger; fatal poisoning can occur even with minimal distress during exposure. The development of lung edema is indicated by burning in the chest, labored breathing, and spasmodic coughing, with pneumonia being a common sequela. Chronic exposure can lead to headache, sleeplessness, loss of appetite and weight, dyspepsia, constipation, mucous membrane ulcers, and dental erosion.

Treatment

Any person suspected of nitrogen oxide exposure requires immediate and complete bed rest for 24 hours, along with the application of warmth to the body. If the respiratory rate is elevated, oxygen should be administered via nasal catheter or, preferably, inhalation. In cases of cyanosis, oxygen concentration should be adjusted to restore the patient's normal color. Cardiac stimulants, blood pressure-increasing drugs, morphine, or barbiturates (which depress the respiratory center) should be avoided.

Aviation Fuels: Toxic Properties and Precautions

Aviation fuels, whether gasoline or jet propellant, contain various toxic components that require careful management.

Aviation Gasoline

Aviation gasoline is a complex mixture of aliphatic and aromatic petroleum hydrocarbons, along with special additives like tetraethyl lead and xylidine, present in varying proportions. Grades and types of aviation fuel used by the USAF include 80, 91/96, 100/130, and 115/145 octane (MIL-F-5572). One gallon of gasoline, upon complete evaporation, generates approximately 30 cubic feet of vapor at sea level; these vapors are denser than air. Given their ready absorption by the pulmonary epithelium, their toxicity is of significant practical importance, with adverse reactions observed in flying personnel exposed to volatilized gasoline. Future aviation gasolines may incorporate xylene and/or toluene in amounts up to 10%, depending on grade and manufacturing processes.

Pharmacology of Aviation Gasoline

The concentration of gasoline vapors tolerable by humans is well below the levels required to form combustible or explosive mixtures with air. High concentrations of gasoline vapor in the air can lead to extremely rapid absorption through the lungs, with symptoms appearing within minutes of exposure. Even one-tenth of the concentration necessary for combustion or explosion can be harmful if inhaled for more than a short period, causing dizziness, nausea, and headache. Larger amounts act as anesthetics, inducing unconsciousness. The maximal safe concentration for exposure to ordinary gasoline vapors is about 500 ppm (0.05%). However, due to its aromatic hydrocarbon content, aviation gasoline is likely at least twice as toxic. Furthermore, given the precise and often complex tasks demanded of flying personnel, even small amounts of gasoline vapors in the aircraft must be considered dangerous.

The detection of fumes can induce psychological excitability which, combined with toxicological excitability, may impair judgment among aircrew. This pharmacological excitability, often overlooked, is a probable cause of accidents attributed to pilot error. As gasoline fumes are not inherently unpleasant, they may not trigger sufficient concern from aircrews. Therefore, it is crucial to emphasize that upon detecting gasoline fumes, aircrew must utilize 100% oxygen to prevent inhalation.

Symptoms of Aviation Gasoline Exposure

Gasoline-induced symptoms and pathological changes result from both its irritant and lipolytic actions. The volatile, aliphatic, saturated hydrocarbons exert physicochemical effects due to their high solubility in fat, accumulating particularly in the lipoid constituents of the nervous system and blood corpuscles, where they cause detrimental effects. Acute poisoning is characterized by burning eyes, lacrimation, severe cerebral symptoms such as restlessness, excitement, disorientation, disorders of speech, vision, and hearing, convulsions, coma, and ultimately death.

Tetraethyl Lead

Tetraethyl lead, an antiknock additive, is highly toxic and can cause poisoning through skin absorption or vapor inhalation. Unlike inorganic lead, tetraethyl lead, being an organic compound, primarily affects the central nervous system in poisoning cases. Symptoms include insomnia, mental irritability, and instability, potentially progressing to lead encephalopathy with acute mania. In less severe cases, sleep disturbances with restlessness and terrifying dreams are noted. Other symptoms encompass nausea, vomiting, muscle weakness, tremor, myalgia, and visual difficulties.

Specification	Grade	Tetraethyl Lead (cc/gal) Maximum	Aromatics
MIL-F-5572	80 Octane 91/96 Octane 100/130 Octane 115/145 Octane	0.5 4.6 4.6 4.6	3-15% 3-15% 3-15% 3-15%
MIL-J-5616	JP-1	0	0-20%
MIL-F-5624a	JP-3 JP-4	0 0	0-25% 0-25%
MIL-3-3056	Motor Vehicle Gasoline	3	Varies by Process of Manufacture
VU-M-561	11	3	(None required by specification)

The concentration of tetraethyl lead in aviation gasoline, approximately 4.6 cc/ gal., is sufficiently low that a lead hazard from normal handling is remote. Poisonings encountered in the Air Force have typically resulted from entering gasoline storage tanks containing concentrated amounts of tetraethyl lead within accumulated sludge, or from maintenance activities such as welding, buffing, and grinding on engines that have used leaded gasolines, which can lead to significant exposure to lead compounds.

JP Fuels

JP fuels used by the Air Force are categorized into three grades: JP-1, JP-3, and JP-4. JP-1 is primarily paraffins, similar to kerosene, and contains up to 20% naturally occurring aromatics. JP-3 is a blend of one-third fuel oil, one-third kerosene, and one-third gasoline, with up to 25% naturally occurring aromatics. JP-4 possesses a narrower distillation range and also contains up to 25% naturally occurring aromatics. Significantly, unlike aviation gasoline, JP fuels do not contain tetraethyl lead.

The recommended threshold limit for JP fuel vapors is 500 ppm. Toxic effects can occur below explosive levels, meaning that a toxicological concern persists even in the absence of an immediate fire hazard. Inhalation of JP fuel vapors can lead to slight narcotic effects, similar to those caused by other hydrocarbon vapors, and may induce conjunctivitis. As JP fuels may contain a higher concentration of toxic aromatics than aviation gasoline, they should be handled with equivalent precautions.

Hydraulic Fluid Vapors: Types and Hazards

Leaks in hydraulic systems can release fluid vapors into aircraft compartments, posing toxicological risks to occupants.

Sources and Types

A minor leak from a hydraulic pipe or gauge under pressure can generate a fine spray of fluid that rapidly disperses throughout the cockpit. Larger leaks may lead to the accumulation of liquid pools on the floor. In either scenario, the cockpit air can quickly become saturated with the volatile constituents of the hydraulic fluid, making their toxicity a relevant concern.

Currently, two primary types of hydraulic fluid are used in the Air Force: Petroleum Base Fluid (MIL-O-5606) and Castor Oil Base Fluid (Spec. No. 3586C). Petroleum Base Fluid is used in virtually all USAF aircraft, while Castor Oil Base Fluid has very limited use, primarily in certain trainer aircraft.

Pharmacology and Hazards

Significant differences in toxicity exist between the two hydraulic fluid types.

Petroleum Base Fluid (MIL-O-5606): This fluid is primarily composed of a mineral oil base, a viscosity index polymer, and 0.5% tricresyl phosphate. These components are characterized by relatively low volatility, and their vapors exhibit low toxicity.

Castor Oil Base Fluid (Spec. No. 3586C): This formulation includes a castor oil base, diacetone, butyl cellosolve, ethylene and propylene glycol, and octyl and isoamyl alcohols in varying proportions. The volatile constituents of this fluid, particularly butyl cellosolve, glycol derivatives, and alcohols, are toxic upon inhalation. The alcohols, for instance, are approximately 12 times more potent as narcotics than ethyl alcohol. They also cause considerable irritation of the eyes and respiratory tract, along with headaches and vertigo. Butyl cellosolve vapors can induce similar effects, including eye and respiratory tract irritation, headaches, vertigo, and impairment of judgment and vision. Experimental animals have succumbed within a few hours following a single exposure to air containing 3 mg/L (approximately 700 ppm) of butyl cellosolve.

The toxic effects from inhaling vapors of Spec. No. 3586C are exacerbated by increasing temperature or altitude, both of which serve to increase the concentration of the vapors. In addition to its higher toxicity, hydraulic fluid No. 3586C has a significantly lower flash point (115-140°F) compared to Petroleum Base Fluid (200°F). This difference, coupled with its higher volatility, means it ignites more easily and propagates flames more rapidly.

Coolant Fluid Vapors and Oil Fumes

Beyond fuels and hydraulic fluids, other operational liquids can generate hazardous vapors and fumes within aircraft.

Coolant Fluid Vapors

Coolant fluid, used in liquid-cooled engines, typically consists of ethylene glycol diluted with varying amounts of water, up to 80%, depending on the specific aircraft type. A small quantity of an inhibitor, NaMBT, is also present at a ratio of about 1 to 2,000. While ethylene glycol is toxic if ingested, its vapors do not exert significant toxic effects through inhalation. Even prolonged exposure (several months) to ethylene glycol vapors has shown no deleterious effects other than moderate irritation of the respiratory passages. To date, no instances of intoxication from coolant fluid vapors in flight have been reported. However, breaks in coolant lines often lead to smoke in the cockpit, either from overheating or from the fluid itself. Smoke in the cockpit is a serious concern for pilots, and in several instances, aircraft have been abandoned due to coolant line leaks. Despite ethylene glycol having a flash point of 117°F, the fire hazard from escaping coolant fluid is not considered high, especially given its diluted state.

Oil Fumes

Aircraft oil hose connections, typically employing adjustable clamps, are prone to breaking or loosening, unlike the pressure-type connections used in hydraulic systems. When oil escapes onto hot engine parts, smoke is frequently formed, subsequently finding its way into the cockpit. As documented by Armstrong, individuals breathing hot fumes during flight have reported symptoms similar to carbon monoxide poisoning, including headache, nausea, and sometimes vomiting, in addition to irritation of the eyes and upper respiratory passages. While the exact chemical compounds responsible are not precisely identified, they are believed to include methyl and ethyl aldehyde, acrolein, and paraformaldehyde, which are principal breakdown products of lubricating oil.

Fire Extinguishants: Toxicity and Effects

Fire extinguishants, vital for safety, also pose potential toxicological risks in enclosed aircraft environments.

Types of Fire Extinguishants

Three primary chemicals are commonly used as fire extinguishants in aircraft. For fixed systems, carbon dioxide (CO_2, Spec. 14069) and chlorobromomethane (CB, Spec. 14163) are employed. Hand extinguishers in aircraft contain either carbon dioxide (2.66 pounds capacity) or chlorobromomethane (1 quart capacity), with CB having replaced carbon tetrachloride in these units.

Carbon Dioxide (CO_2)

The initial physiological effect of inhaling carbon dioxide is typically noted at concentrations around 2%, where breathing becomes labored and total respiratory volume increases. At 4%, the depth of respiration markedly increases, and at 4.5% to 5%, breathing becomes labored and distressing for some individuals. At or near maximum voluntary tolerance, concentrations of 5% to 10% can lead to a failure of compensatory reactions, while concentrations exceeding 10% result in marked deterioration and an inability to take self-preservation steps. Signs and symptoms of CO_2 poisoning include excitement, headache, vertigo, dyspepsia, drowsiness, weakness, dizziness, and muscular weakness. High concentrations can culminate in coma or death.

Carbon Tetrachloride (CC14)

(Note: While chlorobromomethane has largely replaced carbon tetrachloride in hand extinguishers for aircraft, understanding its toxicity remains relevant for historical context or potential encounter with older equipment.)

The lowest detectable concentration of carbon tetrachloride vapor by smell is approximately 72 ppm (0.0072%). However, the maximum allowable safe concentration for an 8-hour daily (chronic) exposure is generally set at 25 ppm (0.0025%). Therefore, if carbon tetrachloride is detectable by odor, the air is unsafe for prolonged exposure. Concentrations exceeding 1,000 ppm (0.1%) can induce symptoms of acute poisoning within minutes.

Acute CC14 intoxication primarily targets the nervous system, exerting a narcotic action. Its secondary, toxic actions involve the kidneys and liver, leading to a hepatorenal syndrome. The kidneys are affected far more frequently than the liver, with renal injury characterized by epithelial destruction and stroma escape. Liver involvement is less common and typically mild.

Acute intoxication is characterized by headache, dizziness, drowsiness, and lassitude. In severe cases, nausea, vomiting, abdominal pain, and diarrhea may occur. Within a few hours post-exposure, urine may become scanty and dark, potentially leading to anuria, accompanied by puffy eyes, edematous ankles, and ultimately coma. Jaundice may appear if the liver is involved. The subsequent course typically resembles a toxic nephrosis, usually followed by recovery, though visual field limitation may persist if the optic nerve was affected. Chronic exposure can cause headache, vertigo, appetite loss, strength reduction, and progressively acute fatigue. Daily skin contact with carbon tetrachloride may also result in a scaling, fissure-like dermatitis.

Chlorobromomethane (CB)

Chlorobromomethane is a narcotic agent of moderate intensity but prolonged duration, emphasizing the importance of avoiding acute exposures. Exposure to high concentrations of CB vapor can cause staggering, uncoordination, stupor, confusion, headache, nausea, and dizziness. Its chronic toxicity is very low, with adverse effects generally not expected below 0.01%.

In contrast to the intensity of its narcotic action, acute exposure to chlorobromomethane is less likely to cause liver necrosis compared to carbon tetrachloride, though it may induce fatty degeneration of the liver. Overall, the chronic toxicity of CB is significantly lower than that of carbon tetrachloride. However, decomposed CB vapor is considerably more toxic than its undecomposed form. For example, the approximate lethal concentration for rats exposed to undecomposed vapor for 15 minutes was 343 mg/L, while for decomposed vapor (at 800°C), it was 22 mg/L. When heated to decomposition, CB emits highly toxic fumes of chlorides and bromides, which are irritating and damaging to the lungs. Accumulations of these fumes in confined spaces, such as aircraft cockpits, can have serious consequences.

High Energy Fuels (Boranes): Specific Toxicities and Safety

High energy fuels, a class of boron-hydrogen compounds, offer a higher heat of combustion than conventional hydrocarbon fuels, yielding greater energy per unit weight. Their utilization in high-performance aircraft significantly enhances range, speed, and overall performance. However, these propellants are characterized by extreme toxicity.

Composition and Toxicity

The most frequently considered boron hydrides for high energy fuels are pentaborane and decaborane. Notably, diborane, a common decomposition product of both, also presents significant toxicological concerns.

Inhalation toxicity studies in laboratory animals have indicated that pentaborane and decaborane primarily act as central nervous system irritants, while diborane is predominantly a pulmonary irritant. Decaborane has also been shown to induce similar central nervous system effects following absorption through intact skin. Based on accumulated toxicological data, the recommended maximum allowable concentrations for industrial environments are: Diborane – 0.1 ppm; Pentaborane – 0.01 ppm; and Decaborane – 0.05 ppm. At these threshold limits, only decaborane is detectable by odor.

Symptoms

Symptoms observed in individuals accidentally exposed to boranes in industrial and laboratory settings, listed in decreasing order of frequency, include dizziness and vertigo, tightness in the chest, headache, cough, drowsiness, nausea, nervousness and restlessness, fatigue and weakness, convulsions, and chills and fever. It is notable that central nervous system effects have been experienced by some individuals exposed to diborane, while pulmonary symptoms have developed in others exposed to pentaborane and decaborane. To date, no fatalities have been attributed to borane intoxication.

Treatment

The management of an individual acutely exposed to high energy fuels should commence with prompt removal from the contaminated area. If a spill has resulted in bodily contamination, the skin must be thoroughly washed with copious amounts of water, and clothing removed and decontaminated similarly. If ingestion has occurred, vomiting should be induced. For inhalation exposures, oxygen administration is indicated. Subsequent symptomatic and supportive therapy will be dictated by the specific pharmacological properties of the borane(s) involved. Currently, no specific antidote or form of therapy is available for borane intoxication.

Prevention

An active preventive medicine program is imperative on bases where high energy fuels are handled. Ground personnel and aircrew members must be thoroughly educated on the toxic nature of these materials, the symptoms of intoxication, and appropriate emergency first aid measures. Atmospheric detection devices have been developed and are crucial for evaluating suspected areas of contamination on the ground and within crew compartments. Personnel entering contaminated areas should be fully protected with fire-retardant coveralls, goggles, neoprene gloves, and boots. Respiratory protection is best achieved with an air-line respirator or a self-contained breathing apparatus like the MSA Chemox (rebreather type).

References

USAF. (2022). "Toxicological Information Center (AFR 160-100)". U.S. Air Force. USAF. (2024). "Potentially Toxic Agent Precautions (AFR 160-108)". U.S. Air Force.

USAF. (2021). "Engineering Data, Preventive Medicine and Occupational Health Program (AFM 160-25)". U.S. Air Force.

Armstrong, H. G. (2018). Toxicology in Aviation. In "Aerospace Medicine" (Chapter 25). Williams and Wilkins Co.

Lowe, H. J., & Freeman, G. (2017). Borane Intoxication in Humans. "AMA Archives of Industrial Health", 16, 523-533.

Patty, F. A. (2019). "Industrial Hygiene and Toxicology: Toxicology" (Vol. II). Interscience Publishers Inc.

Headquarters Air Material Command. (2020). "Toxic Hazards in Military Flying and the Aviation Industry: Aviation Medicine Symposium".

Committee on Aviation Toxicology, Aeromedical Association. (2015). "Aviation Toxicology". The Blakiston Company.

Chapter 12: Nutritional Guidelines for Aircrew and Passengers

Nutrition constitutes a foundational element of health and operational readiness for all military forces, holding particular relevance for the United States Air Force (USAF) due to the distinct demands it places on flight requirements for aircrews. The strategic and operational goals of nutritional management within the USAF are addressed through a tripartite framework: the Ground Feeding Program, which caters to personnel stationed at Air Force Bases; the Flight Feeding Program, a specialized initiative designed for airborne situations; and the Survival Feeding Program, conceived to provide essential sustenance for airmen isolated in challenging or remote environments. This chapter delves into each of these programs, outlining the guidelines, challenges, and solutions for maintaining optimal nutritional status among flying personnel and passengers.

Fundamentals of Aircrew Nutrition

The overarching principle of aircrew nutrition is to ensure that personnel maintain optimal physical and cognitive function, critical for the complex and demanding nature of aerospace operations. Flying activities inherently disrupt and modify fundamental living habits, including sleep, eating, and drinking patterns. Therefore, a primary objective of flight feeding efforts is to facilitate the adjustment of both aircrews and aircraft passengers to these unique work demands. Field observations consistently indicate that irregular eating practices, or "nonfeeding," sustained over extended periods, contribute significantly to fatigue, increase the likelihood of human error, and may even lead to aircraft accidents. Beyond operational safety, the value of flight feeding is widely recognized for its positive impact on general bodily comfort and morale. To promote peak performance, the flight feeding system is meticulously designed to "refuel" human operators with necessary nutrients, administered on a careful and regular schedule, mirroring the precision with which an aircraft is refueled.

Ground Feeding Programs

The ground feeding program for Air Force personnel is meticulously managed within base dining halls or cafeterias, operating under a system aligned with that of the U.S. Army. Under this established system, all ration supplies are centrally procured and distributed through Army quartermaster channels. This process is orchestrated based on a comprehensive variety of complete meal menus, which are circulated months in advance by a Joint Army-Air Force Master Menu Board. These master menus are diligently planned and executed in strict adherence to the nutritional standards prescribed within AFR 160-95. The nutritional integrity of these dietary standards must be rigorously maintained, even when subsequent adjustments to the menus become necessary due to local climatic conditions, personnel requirements, or supply chain dynamics.

Final localized modifications to the meal menus are formally authorized and are typically coordinated through base menu boards. These boards comprise key representatives, including the food service supervisor, the commissary officer, and the base surgeon, ensuring a holistic approach to dietary planning. This intricate arrangement is specifically designed to guarantee satisfactory ground feeding practices, despite the inherent complexities and global operational scope characteristic of Air Force endeavors.

The primary standard ration designated for ground use is the Field Ration A. This ration is typically issued to Air Force units that possess both kitchen and refrigeration facilities. It encompasses a wide array of fresh, perishable food components, meticulously listed within the master menus and regularly served at bases situated within the Zone of Interior. In situations where such perishable items cannot be readily stocked at overseas or remote field locations—often due to a lack of adequate refrigeration facilities—the operational B Ration is provided as the standard dining hall ration. The B Ration substitutes perishable items with canned or dehydrated equivalents, comprising nonperishable variants of the same food types found in the Field Ration A, designed to feed groups of approximately fifty or more individuals.

For smaller Air Force units operating without immediate access to kitchen facilities for temporary periods, adequate sustenance can be maintained through the Ration, Small Detachment, 5-in-1. Each packaged ration of this type is formulated to provide food for five individuals for one day, and it can be consumed either hot or cold. Its utilization within the Air Force is generally restricted to emergency reserve provisions for advanced radar and weather detachments, crash crews, and search and rescue operations.

Additional ground-type packaged rations or specialized supplements are comprehensively enumerated and detailed in T.O. No. 00-35A-36, titled "Operational Rations, Food Packets, and Supplements." This inventory includes Ration Supplements specifically tailored for hospitals or aid stations, the Ration, Individual, Combat, and the Ration Arctic Trail. The latter two rations are primarily conceptualized for Army field forces operating under combat conditions. They are specifically engineered to furnish food for one man for a single day and are therefore applicable with relative infrequency to the specific requirements of the Air Force.

Flight Feeding: Pre-Flight Preparations

Flight feeding is conceptualized across three distinct yet interconnected categories: pre-flight, in-flight, and post-flight. These categories represent specialized extensions to the fundamental ground nutrition program, necessitated by the progressively extended ranges and enhanced performance capabilities of modern aircraft in recent years.

Effective pre-flight preparations are paramount, mandating that each individual embarking on an aircraft should consume a freshly prepared, balanced meal approximately one to two hours prior to the anticipated take-off. This meal typically aligns with a breakfast menu, comprising relatively light proportions, irrespective of whether it is scheduled during the day or night. The objective is to foster a desirable state of relaxation and promote regularity of digestion through consumption in pleasant, unhurried eating conditions.

For fighter pilots and certain bomber crews, stricter dietary control may be required to mitigate the incidence of gas pains and enhance crew effectiveness, particularly at high altitudes. However, the implementation of specific fixed diets is often not entirely satisfactory due to the notable variability in food tolerances and preferences among individuals. Generally acceptable meals are characterized by a high carbohydrate content and are devoid of foods known to produce flatulence or excessive bulk in the colon.

Foods expressly contraindicated for pre-flight consumption, owing to their propensity to induce abdominal gas, include vegetables from the cabbage family, dried peas and beans, carbonated beverages (including beer), turnips, rutabagas, and other fibrous raw fruits or vegetables. The chewing of gum is also discouraged as it is recognized to promote air swallowing. Conversely, many fresh fruits and fruit juices are permissible and actively encouraged, as they contribute to the prevention of vitamin C depletion, a concern with repeated altitude exposure. Food items that are high in fat, heavily spiced, or improperly cooked are less readily digested and are generally avoided by aircrews. Field reports substantiate that the occurrence and severity of gastric distress during flight are considerably low when moderate dietary precautions are diligently observed.

Alert-crew feeding constitutes a specialized aspect of pre-flight nutrition. Personnel assigned to alert crew status are confined to the alert crew hangar and must maintain a continuous state of readiness for immediate takeoff. AFR 146-16 grants local commanders the authority to establish dedicated special dining facilities to cater to this specific situation, encompassing both pre-flight and post-flight meals. Food items authorized for alert crew feeding include those specified by AFR 145-11, alongside precooked frozen meals and individual in-flight (IF) food packets.

In-Flight Nutrition: Challenges and Meal Types

In-flight feeding represents a relatively recent development compared to other established aircraft procedures. Historically, early aircraft had shorter flight durations, obviating the necessity for organized in-air feeding. However, the critical importance and demand for such provisions became unequivocally apparent during World War II. The contemporary understanding and practices of in-flight nutrition have progressively evolved in response to the escalating requirements of aircrews for extended missions, a direct consequence of ongoing developments in range-extension technologies. Consequently, a certain degree of in-flight feeding is now standard practice across most Air Force operations.

The success and scope of food servicing in an aircraft are influenced by a myriad of factors. Meals consumed aloft often represent a nutritional compromise, dictated by the practical realities of confined aircraft space, limited equipment, and the operational demands inherent in the flying situation. As such, no single method of in-flight feeding or a solitary standardized type of food packaging can comprehensively satisfy the dynamic and varied needs of flyers. A truly satisfactory feeding operation must prioritize simplicity, ease of logistical support, and offer a diverse selection of well-liked foods and beverages presented in appealing combinations. To some extent, this necessitates a distinct "prescription" of meal types and food servicing equipment tailored to each specific aircraft model and the unique requirements of each flight mission.

Air Force equipment directives stipulate that drinking fluids must be supplied in all aircraft capable of sustained flight exceeding three hours. These fluids are provided in quantities of one quart per crew member or passenger for every sixteen hours of flight. Similarly, provisions for flight lunch storage and heating facilities are scheduled for aircraft with flight durations exceeding six hours, with an allowance of one additional meal for each subsequent six-hour interval. This criterion serves as a foundational guide for the initial authorization, design, and production of feeding apparatus and food packets.

However, these planning figures exhibit considerable flexibility, as actual feeding practices are refined within the operational Air Force commands. For instance, the sheer aircraft flight time has not proven to be the sole accurate index for determining in-flight feeding requirements. For this specific purpose, "flight duration" should encompass the total time elapsed from the pre-flight breakfast (or the last meal consumed before takeoff) until the conclusion of post-flight debriefing or interrogation.

Field observations have revealed consistent trends in aircrew feeding habits that are prevalent across various Air Force commands, often categorized as "in-flight peculiarities." A notable observation is the general decrease in the appetites of crew personnel, particularly during the culminating hours of extended flights, leading to a more critical perception of food items. Intrinsic characteristics of the military aircraft environment, including the intense concentration required for work, pervasive noise, continuous vibration, and reduced oxygen availability, collectively contribute to a diminution of digestive processes. In instances of extreme tension arising from air emergencies or active combat, gastric function may be completely inhibited.

The palatability of certain food items can vary significantly between ground level and altitude, beyond the influences of diminished appetite, excitement, or fatigue stress. Comparative acceptance studies have indicated that foods such as potatoes, vegetables, and salads are rated approximately 20 percent lower in acceptability when consumed in the air compared to on the ground. Conversely, soups, meats, fruits, and beverages generally demonstrate comparable acceptance in both environments. Baked goods and desserts, however, are consistently perceived as highly palatable across all flight circumstances.

Dietary monotony poses an additional in-flight challenge specifically for aircrews, though it is less pertinent to airborne troops and passengers who travel less frequently. Passenger personnel generally tend to consume more substantial meals, presumably as a coping mechanism to alleviate flight strain or tedium. This eating behavior also serves to prevent feelings of "emptiness" and other forms of gastric discomfort that are believed to predispose certain susceptible individuals to airsickness.

A recommended interval of approximately six hours is advised between in-flight meals (AFR 146-16). Nevertheless, the provision of small quantities of "free choice," sugar-yielding food supplements is considered desirable during the intervals between main meal periods. Beverages are of paramount importance and should be freely and continuously available. These aforementioned factors are presented as guiding principles for the average in-flight practices observed among the majority of operational personnel, rather than as rigid and inflexible requirements.

To mitigate excessive repetition in meal offerings and the consequent decrease in acceptability, seven distinct types of flight meals have been formally authorized. The utilization of any other meal types is strictly contingent upon authorization from Headquarters USAF.

Authorized Meal Types

Flight Meals Authorized for General Use: These meals are designed for interchangeable use, contingent upon flying schedules, specialized equipment availability, and mission parameters.

1. **Food Packet, Individual, In-Flight (IF):** This packet contains canned items and is specifically engineered for deployment from bases where fresh foods are unavailable or where their storage in aircraft without spoilage is impractical. Each packet functions as a complete meal, with ten distinct menus assembled into separate units. These packets boast a storage stability of approximately two years. Each in-flight food packet typically provides an average of 1,200 calories and has demonstrated high acceptability, particularly when consumed at irregular intervals.

   Each food packet contains four cans: one meat item, one fruit item, one bread component, and one dessert unit. The ten menus encompass five varieties of fruit, eleven distinct meat items, and five dessert selections.

   Specific components include: * **Meats:** Beef steaks, Beef and Corn, Chicken, Ground meat, Chicken and Noodles, spaghetti, Ham and eggs, Pork Steaks, Fried Ham, Turkey, Hamburgers, Tuna. * **Fruits:** Apricots, Pears, Fruit Cocktail, Pineapple, Peaches. * **Desserts:** Cookies, Pound cake, Fruitcake, Date pudding, Pecan roll. * **Bread:** Provided.

   Additionally, each menu features an accessory packet containing individual servings of soluble cream, coffee and tea, sugar, and chewing gum. While all food items are pre-cooked and edible cold, the flavor profile of the meat items and date pudding is significantly enhanced through heating. Several types of food warming devices have been authorized for installation and use on aircraft. This food packet is lauded as the most versatile in-flight meal option due to its nonperishable nature, broad applicability across the majority of aircraft types, ready availability on short notice through established supply channels (commissaries, personal equipment offices, and flight kitchens), and its minimal requirement for aircraft servicing equipment.

2. **Precooked Frozen Meal:** The main dishes for the precooked frozen meal program are centrally procured and distributed via commissary supply channels. Complementary items, such such as bread, salad, a beverage, and dessert, are issued to the crewman by the flight kitchen. A total of 12 distinct menus are available in this category.

These menus include: * **Menu 1:** Turkey, lima beans, mashed sweet potatoes. * **Menu 2:** Swiss steak, peas, au gratin potatoes. * **Menu 3:** Beef patty, green beans, mashed potatoes. * **Menu 4:** Chicken breast and thigh, corn, oven-browned potatoes. * **Menu 5:** Beef pot roast, mixed vegetables, oven-browned potatoes. * **Menu 6:** Tenderloin steak, green beans, mashed potatoes. * **Menu 7:** Chicken pot pie. * **Menu 8:** Beef pot pie. * **Menu 9:** Waffles, veal sausage patty, applesauce. * **Menu 10:** Omelet, veal sausage patty, sweet roll. * **Menu 11:** Breakfast steak, cottage-fried potatoes, fruit compote. * **Menu 12:** French fried chicken, steamed rice, stewed apricots.

Menus 1 through 8 are designated as dinner meals, while Menus 9 and 10 serve as breakfast options. Menus 11 and 12 offer flexibility for either breakfast or luncheon. The storage and preparation of all these menus necessitate the presence of aircraft ovens and refrigerators. These meals are procured quarterly, and their cartons are marked with the date of production. For optimal acceptability, they should ideally be consumed within 9 months from the date of manufacture. A critical safety feature is a plastic vial, half-filled with water and frozen in a vertical position, placed in each case in a lateral position. If the ice within this vial has melted and flowed along the axis of the tube, the meals are deemed unsafe for consumption. The melted ice serves as unequivocal evidence that the temperature has been sufficiently elevated to potentially allow the presence of Staphylococcal toxins.

3. **Sandwich Meal:** The Sandwich Meal is by far the most prevalent type of in-flight meal, prepared as a standard Air Force package either in dining halls or by specialized flight kitchens. It comprises fresh sandwiches, milk, canned juices, fresh fruit or desserts, supplemented by additional items such as celery, pickles, and hard-boiled eggs. Various nutritionally balanced combinations are detailed in AFR 146-16. However, their essential freshness, visual appeal, and appetite-stimulating qualities are largely dependent on the resourcefulness and diligence of the kitchen personnel.

The most acceptable sandwiches are those containing sliced meats, chicken, or turkey, from which all bone fragments, bone splinters, and inedible gristle have been meticulously removed. These sandwiches must be wrapped immediately after preparation in waxed paper sandwich bags and subsequently refrigerated below 40°F until they are issued to crew members. The inclusion of gravy, chopped egg, or chopped meat fillings is strictly prohibited due to the significantly increased risk of bacterial food poisoning associated with these items.

After five hours or more at ambient room temperatures, most types of sandwiches and other perishable items can become unsafe for consumption due to the production of toxins resulting from bacterial growth. Consequently, all sandwich components not consumed within five hours of preparation or issue must be destroyed. The sandwich lunch proves most useful in situations where no installed aircraft equipment is available and is generally well-received, provided it is not offered with excessive frequency. Practically, its use is typically limited to short flights or as the initial meal during extended missions.

Flight Meals Authorized for Specific Use:

1. **Bite-Size Meal:** The bite-size meal is authorized specifically for jet aircraft operations where the serving of any other type of flight meal is impractical. All components of this meal must be of a "bite-size" dimension and suitable for consumption by hand. Each package is clearly marked with the date and the time limit for safe consumption, requiring consumption no later than five hours after preparation. The bite-size meal comprises the following components:

   o   Beverage Unit: Milk or juice.

   o   Meat Component: Cubes of cooked steak or other lean, tender meats.

   o   Dessert Component: Cookies or pieces of fruit, and candy.

   o   Optional Items: Gum, relishes, nuts, coffee.

2. **Foil-Pack Meal:** The foil-pack feeding system has been specifically authorized for use at certain bases that support particular types of operations, such as radar picket patrol missions. This system is primarily designed for deployment in large aircraft where ample space and power resources are available, and where weight is not a limiting factor. The Strategic Air Command pioneered and first demonstrated the significant possibilities of this procedure, utilizing the Type B-4 oven, which was originally supplied with B-36 aircraft. In their preliminary trials, numerous fresh-chilled food ingredients were prepared and cooked with notable success within hand-assembled, aluminum-foil packages.

   The current iteration of the foil-pack meal consists of five distinct menu items, each contained in separate units: a meat component, two vegetable components (typically potato and another vegetable), a hot roll, and a dessert. Four breakfast menus are available from a comprehensive total of sixty-eight menus that have been developed. With the exception of the packaging of rolls, pies, and cakes, and the initial searing of meat in ground kitchens, all items are packaged uncooked in separate, rectangular foil containers. These containers are then hermetically sealed with a top cover, assembled as a complete meal on single trays, and refrigerated (at 37°F) until the time of final cooking.

   The system relies on three specialized articles of equipment: aluminum foil packs and a crimp closure device, an aircraft refrigerator, and an oven. These meals, prepared with remarkable simplicity in ground support kitchens, are composed of common, lower-cost subsistence supplies typically found in dining halls. They require minimal training and effort from aircrews. This meal concept boasts very high acceptability and is exceedingly popular at installations where it is utilized.

Hydration and Beverage Management

The critical importance of proper hydration for aircrew and passengers cannot be overstated. Dehydration of the human body leads directly to lowered efficiency, posing a serious threat to flight operations, particularly in hot climates or at high altitudes. Therefore, a robust hydration strategy is an integral part of nutritional guidelines.

A variety of beverages are deemed popular and suitable for consumption in flight, including cool water, coffee, tea, chocolate milk, tomato juice, and various fruit juices. Cool water is considered essential and should always be readily available. For missions extending beyond a few hours, other aforementioned beverages should also be made freely accessible. Furthermore, beverages are an expected and necessary component to be included with all flight meals. However, it is important to note that the gratuitous issue of government-provided beverages to passengers and crews between structured meal periods is not authorized.

To facilitate effective hydration, specific liquid feeding equipment is available. One- or two-gallon containers are typically employed in passenger, cargo, and bomber aircraft, where a significant number of people must be served and where crew member mobility within the cabin is permitted. The recommended two-gallon capacity container is designated as the "Jug, Insulated, type CNU-2/ C" (standard, specification MIL-J-25718). This unit is rectangular in shape, constructed from stainless steel, and features an electrical element capable of operating on either 28 Volts DC or 115 Volts AC. Its design ensures that liquids are maintained within a temperature range of 170°F to 190°F as long as electrical power is supplied. This jug can also be charged with wet ice to effectively keep beverages cool. With an initial full charge of ice, and at an ambient temperature of 90°F, the liquid temperature within the jug can be sustained below

45°F for a period ranging from 16 to 25 hours. The type CNU-2/C jug has superseded the Type J-1 container, which featured a dry ice well and is now considered limited standard.

An alternative two-gallon container, suitable for use when electrical power is unavailable or when a cylindrical shape is preferred, is the Type III, Grade A, Class 2 Insulated Jug, described by specification MIL-C-3164. This container is also constructed of stainless steel and is available in both one-gallon and two-gallon sizes. These jugs are designed to maintain beverages above an acceptably warm temperature for a minimum period of six hours at an ambient temperature of 68-76°F. This type of jug replaces the older Type F-1 liquid containers, which are now considered limited standard.

For fighter aircraft and situations requiring a crew member to remain in a fixed position for an extended duration, special equipment is provided. Crew position water bottle assemblies, available in one-quart horizontally installed and two-quart vertically installed types, are now standard issue items. Each assembly comprises a stainless steel vacuum bottle equipped with a cap, a sealing gasket, and a spigot. The spigot incorporates a vent tube designed to allow liquid to drain efficiently from the bottle. The liquid outlet of the spigot is connected to a length of silicone rubber tubing, which terminates in a Teflon drinking probe. A handset valve is used to regulate the flow of fluid. The container is mounted such that the liquid flows by gravity to the point of consumption. These bottles are engineered to keep liquids above an acceptably warm temperature for at least six hours at an ambient temperature of 77°F.

Furthermore, a device has been developed for piercing commercial juice and beverage cans, enabling direct consumption of the liquid from the can. This device is included in the "Dispensing Kit Liquid Can Piercing-Drinking," described by USAF Dwg. No. 54B3827. While this specific item is currently standard issue, it remains under ongoing development and modification to enhance its functionality and user experience.

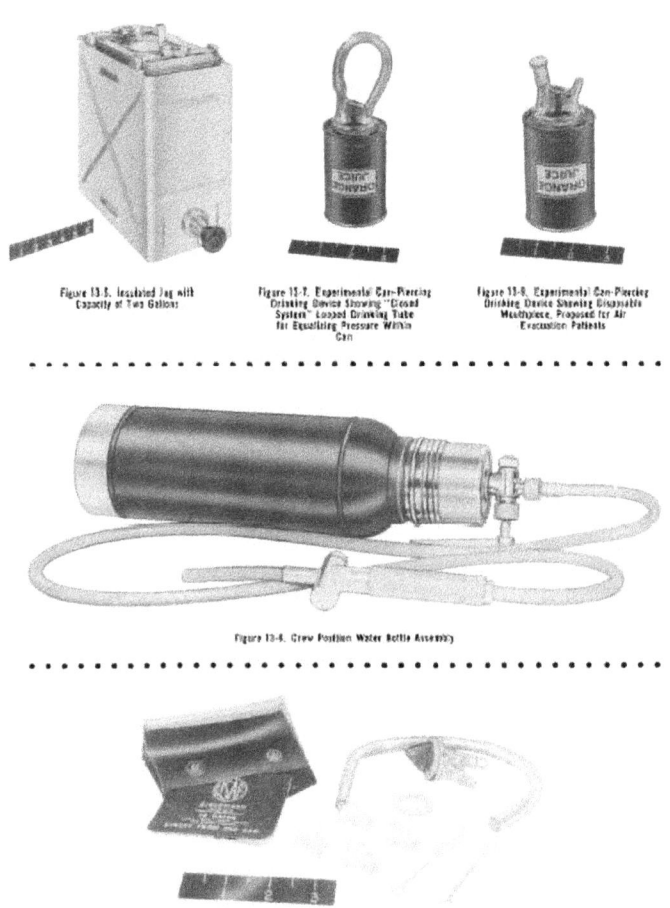

Figure 13-5. Insulated Jug with Capacity of Two Gallons

Figure 13-7. Experimental Can-Piercing Drinking Device Showing "Closed System" Looped Drawing Tube for Equalizing Pressure Within Can

Figure 13-8. Experimental Can-Piercing Drinking Device Showing Disposable Mouthpiece, Proposed for Air Evacuation Patients

Figure 13-6. Crew Position Water Bottle Assembly

Food Servicing Equipment and Galleys

The efficacy of flight feeding is intrinsically linked to the availability and functionality of specialized food servicing equipment. Details of completed developmental studies concerning feeding procedures and equipment are comprehensively documented in Technical Memorandum Reports distributed by the Wright Air Development Center. Several projects of particular common interest are pertinent to this discussion, including fighter in-flight feeding, the foil-pack feeding system, general food servicing equipment, and the microbiology of flight meals.

Fighter In-Flight Feeding: This is not typically considered desirable or required due to the generally short duration of jet-fighter operations. However, the implementation of air refueling or other range-extension techniques necessitates the provision of cockpit feeding. This prospect demands serious consideration, even if its application is limited to a few fighter aircraft models and a constrained number of missions. Preliminary studies suggest that fighter in-flight feeding can be satisfactorily achieved through a reduced caloric intake, predominantly comprising liquids dispensed via specialized apparatus. Further conclusions are reserved pending clearer definition of firm operational requirements for long-range fighters. It is anticipated that ongoing experimentation will yield tested procedures readily available for selection as Air Force needs emerge.

Food Servicing Equipment: While briefly referenced, a multitude of additional items exist, either as recent developments or previously standardized components for aircraft supply. The proportion of in-flight feeding problems and deficiencies should diminish commensurately with the ready availability of such apparatus to aircrews. Principal types of equipment items include various heating and cooling units, dispensing kits, and specialized containers.

The **B-4 in-flight feeding oven** is designed for heating precooked frozen meals, foil-pack meals, and individual in-flight (IF) canned components in aircraft utilizing 28 volt DC, 120 volt single-phase AC, and 208 volt three-phase AC power. This oven features six removable shelves, each equipped with a 375-watt heating element, which can be re-spaced or heated independently. An additional 175-watt heating element is integrated into the side wall to maintain foods at a warm temperature of 150-160°F. The maximum power draw of the oven is 2425 watts, and it weighs 21 lbs. This oven is capable of warming six precooked frozen meals, six foil-pack meals, or 18 IF canned meat components within approximately thirty minutes. Further information is available in specification MIL-O-6438B (USAF) and technical orders 13B1-2-1 and 13B1-2-4. The B-4 oven is currently being replaced by a newer, forced-air oven boasting superior design and versatility.

The **Type B-3 oven** is designed specifically for warming canned IF food packet components and cans of "ready-to-serve" type soups in aircraft operating on either 28 volt DC or 115 volt single-phase AC power. The oven has a capacity for eight cans, which can be heated to palatable temperatures within 10 to 20 minutes, depending on their initial temperature. The Type B-3 oven has a total wattage of 920, with one 400-watt element embedded in each of two shelves, and a 120-watt "holding" element situated in a side wall. The maximum weight of this unit is 8.5 pounds. The Type B-4 oven, as previously described, is recommended over the Type B-3 due to its enhanced versatility in heating diverse meal types.

Aircraft Galleys: Aircraft food galleys are essentially frameworks that integrate storage space, a work surface, and various items of insert equipment. Specifications MIL-G-25608A and MIL-G-25607 respectively govern the design and testing of these galleys. Current practice involves designing a unique galley for each aircraft type, tailored to the physical space available and the specific feeding requirements of that aircraft.

MIL-G-25608A recommends that insert equipment be selected from the following list: * Rectangular liquid containers in accordance with MIL-J-25718. * Type B-4 ovens in accordance with MIL-O-6438. * Hot cup brackets in accordance with MIL-B-7525, MIL-B-7526, MIL-B-7527, or

MIL-B-7528 (Hot cup brackets may be designed into the galley with approval from the procuring activity). * Hot cups in accordance with MIL-C-7561 and/or MIL-C-7615. *

Drinking cup dispensers. * Refuse container and disposal facilities. * Swing-a-way type, or equivalent, can opener. * Refrigerator (mechanical, dry ice, or other approved type). * Other insert equipment approved by the procuring activity.

Bracket and Receptacle, Hot Cup, four Unit, 28 volts, Type A-1 Spec MIL-B-7528 (Standard)
Bracket and Receptacle, Hot Cup, Single Unit, 28 volts, Type A-2 Spec MIL-B-7526 (Standard)
Bracket and Receptacle, Hot Cup, four Unit, 115 volts, Type B-1 Spec MIL-8-7527 (Standard)
Bracket and Receptacle, Hot Cup, SingleUnit, 115 volts, Type B-2 Spec MIL-8-7527 (Standard)
Bracket and Receptacle, Hot Cup, 115 volts, Single Unit, Type B-2 Spec MIL-8-7525 (Standard)
Cup, food Warming, Electrically Heated, Aircraft, Type A-1 28 volts, Spec MIL-C-7615 (Standard)
Cup, Food Warming, Electrically Heated, Aircraft, Type 8-1 115 volts, Spec MIL-C-7561 (Standard)
Cups and Lids, Paper, Hot Food or Drink, Style A, 603 Spec UU-C-8344 (Commercial Standard)
Dispenser, Paper Drinking Cup, Wall Mounted, Aircraft, 24 Cup Capacity (USAF Dwg No. 4903786) (Experimental)
Dispensing Kit, Liquid Can Piercing, Drinking (USAF Dwg No. 54B3827) (Standard)
Jug, Insulated, Type CNU-2/c (2 Gal.) Spec MIL-J-25718 (Standard)
Jug, Insulated, Type 111, Grade A, Class 2, I Gal. and 2 Gal. Spec MIL-C-3164A (Commercial Standard)
Oven, Food Warming, Electrically Heated, Type B-4 Spec MIL-0-6438B (Standard)
Refrigerator, Dry Ice, Precooked Frozen Food Storage, Type B-1, Weber Aircraft Corp., Burbank, Cal., Dwg. No. R72202 (Com. Stand.) I
Refrigerator, Mechanical, Non Frozen Storage, 4 cu fl, Model SR-4. Dale Sales, Inc., Los Angeles, Cal. (Com. Stand.)
Refrigerator, Mechanical, Non Frozen Storage, 6 cu ft, Model SR-6, Dale Sales Inc., Los Angeles (Commercial Stand.)
Refrigerator, Mechanical, Non Frozen Storage, 12 cu ft, Model SR-GA, Dale Sales Inc., Los Angeles (Commercial Stand.)
Refrigerator, Mechanical, Frozen and Non Frozen Storage, 10 cu ft, Model SR-10, Dale Sales l11c., Los Angeles (Commercial Stand.)
Tray, lnflight, Food Servicing, Disposable Spet MIL-T-8166 (Commercial Standard)
Water Bottle Assemblies, Crew Position, 2 qt Horizontal, 2 qt Vertical, I qt Horizontal, I qt Vertical Spec MIL-8-25337 (Standard)

A water tank, sink or drainage part, and accessory plumbing may also be integrated into the galley design.

Hot Cups and Brackets: The Type A-1 hot cup is designed for operation on 28-volt DC power, while the Type B-1 hot cup is engineered for 115-volt AC power. Both types of cups possess a capacity of 37 fluid ounces. They are designed to provide hot water for reconstituting beverage concentrates, for heating two unopened 211x304 single-strength soup cans or three 300x200 IF ration cans in boiling water, and for directly warming liquid and semi-solid foods. Due to the challenges aircraft facilities often present in adequately cleaning food solids from the cup, its use for substances other than water is generally not recommended. When filled to the brim with water at 70°F (at an ambient temperature of 77°F), these cups are designed to heat the water to 212°F within ten minutes. Both one-unit and four-unit brackets, complete with receptacles, timers, and warning lights, are available for both the 28-volt and 115-volt cups.

· · · · · · · · · · · · · · · · · · ·

Mechanical Refrigerators: The Type C-1 mechanical sectional refrigerator has been superseded by mechanical models offering improved operating characteristics. Several models manufactured by Dale Sales, Inc., Los Angeles, California, are currently in use within the Air Force. * **Model SR-4** is a 4-cubic foot refrigerator engineered to maintain food temperatures within the range of 32-45°F. It includes a small ice cube compartment. Its external dimensions are 34-3/4" high x 24" wide x 24" deep. * **Model SR-6** boasts a volume of 6 cubic feet. It is capable of maintaining food within the 32-45°F temperature range and can accommodate 52 foil-pack or precooked frozen meals (provided conditions permit thawing of the frozen meals). A forced air circulation system ensures rapid temperature pull-down and even temperature distribution throughout the unit. This specific model does not feature an ice cube compartment. Its external dimensions are 33" high x 27" wide x 18-3/4" deep. A refrigeration unit, measuring 12-9/16" wide x 20-1/2" high x 18-3/4" deep, extends from either the left, right side, or the rear panel. * **Model SR-6A** has a volume of 12 cubic feet and is constructed from a basic SF-6 unit augmented by a stack-on section of equal volume. This refrigerator is designed to maintain an internal temperature range of 32-45°F and can hold 104 foil-pack or precooked frozen meals (contingent on conditions allowing thawing of frozen meals). The SR-6A shares the same rapid temperature pull-down characteristics as the Model SR-6. Its external dimensions are 63" high x 27" wide x 18-3/4" deep. It also incorporates an additional refrigeration side unit, similar to that of the Model SR-6. * Model SR-10 is a dual-temperature refrigerator. Its six-cubic-foot upper chamber can be regulated to either +40°F or -10°F and is capable of holding 126 precooked frozen or 98 foil-pack meals. The lower four-cubic-foot section is adjusted exclusively for 40°F and is intended for the storage of items such as milk, butter, fruits, and bread. Its external dimensions are 58" high x 24" wide x 24" deep. A refrigeration unit, measuring 25" high x 24" wide x 11" deep, connects to the box on either the right side, left side, or the rear panel.

Dry Ice Refrigerator: The Type B-1 refrigerator is an insulated aluminum box designed to hold 60 pounds of dry ice in a central well, alongside 32 frozen meals on its sides. When packed in this configuration, it can maintain the meals between 0°F and 20°F for 48 hours, even at an outside ambient temperature of 90°F.

Minor Items: A disposable pasteboard tray has been specifically engineered to accommodate IF cans, foil-pack, and precooked frozen meals. Packet, Accessory, In-Flight Feeding, Type I, constitutes an accessory cellophane packet designed for use with precooked frozen and foil-pack meals, containing a plastic knife, fork, and spoon, a salt envelope, a pepper envelope, and paper napkins. The Type II Packet is intended for use with sandwich snack meals, comprising a plastic spoon, a salt envelope, a pepper envelope, and a paper napkin. Some work has been dedicated to the development of a disposable refuse container. Most galleys are now outfitted with a metal refuse container. It has been determined that satisfactory watertight, disposable inner liners for these refuse containers can be fabricated from polyethylene tubing, cut into appropriate lengths, and heat-sealed on one end.

Microbiology and Safety of Flight Meals

Food-borne infections pose a significant concern and become particularly critical when symptoms manifest during flight. These incidents can occur whenever perishable components of pre-flight and in-flight meals are improperly handled. Therefore, a continuous and stringent preventive control strategy is indispensable. This strategy encompasses meticulous ground kitchen sanitation, rigorous refrigerated storage of packaged in-flight meals whether on the flight line or aboard aircraft, and careful attention to time-temperature factors influencing bacterial growth and the design, use, and cleaning of all servicing equipment.

Microbiological studies clearly delineate the approximate temperature range of 50-130°F as the "danger zone" where food infection organisms rapidly multiply, and enterotoxins can be produced by microorganisms. The minimum incubation period for bacterial growth to reach hazardous proportions is generally established at five hours. Ensuring a safe supply of perishable flight foods consequently mandates adherence to several critical principles: 1. Sanitary Practices: Implementation of robust sanitary practices is essential to prevent the inoculation of pathogens during the ground stages of food preparation and to minimize the total bacterial contaminants. 2. Temperature Control during Holding: Perishable foods should not be held at incubation temperatures (above 50°F) for periods exceeding five hours prior to consumption. 3. Temperature Control during Storage/Serving: Maximum utilization of refrigeration (below 50°F) or, alternatively, heating to above 130°F for continuous periods before serving is required.

These principles apply uniformly to all types of in-flight perishable meal items, including sandwiches, snack lunches, and hot meals, regardless of their origin (flight kitchen, commercial caterer, or household supplies). Individual packaging in disposable, sanitized containers is a highly desirable supplementary protective measure, especially given the limited hygienic facilities inherent in military aircraft.

Repeated bacteriological analyses conducted on perishable in-flight foods, particularly the more complex precooked frozen meals and foil-pack meals, have consistently indicated sufficiently low bacterial counts. This suggests a minimal hazard risk in such feeding, provided that carefully organized supply procedures are strictly followed. Aircraft food heating equipment typically achieves temperatures exceeding 165°F, which effectively inhibits and often destroys food bacteria of pathogenic significance. However, it is crucial to understand that such high temperatures will not inactivate enterotoxins that are more stable, if these toxins have already formed in the food prior to heating. Therefore, complete cooling or freezing is an essential requirement for all protracted periods of food transport and storage. The establishment of consistent bacterial safeguards significantly influences the types of perishable foods that can be safely utilized in aircraft. This is ultimately dependent upon the dedicated efforts and specialized training of personnel directly responsible for the conduct of flight feeding within the operational commands.

Post-Flight and Survival Feeding Strategies

Post-Flight Feeding

The post-flight phase of nutritional support is significantly shaped by the physical and mental condition of returning airmen, which in turn is influenced by the operational and nutritional demands of the completed flight period. Post-flight feeding serves a dual purpose: it actively stimulates physiological recovery processes and simultaneously boosts morale. This strategic approach contributes to shortening the time lost between missions and acts as a crucial preventative measure against the onset of chronic fatigue. For these compelling reasons, post-flight feeding should not be unduly delayed, making convenient flight-line kitchen facilities an absolute requisite.

A key objective of eating during this period is to facilitate the relaxation of tensions induced by prolonged hours of alert concentration or other fatiguing flight pressures. In extreme cases, special provision of light refreshments—such as beverages, ice cream, or juices—may be warranted either before or during post-flight duties like interrogations. This would serve as a preliminary measure before a more complete dinner meal, which should predominantly feature protein. In essence, some degree of feeding is routinely indicated as the initial step in the process of rest and recuperation.

Survival Feeding Strategies

Survival situations refer to emergencies such as bailout, ditching, or other forced landings into primitive, isolated regions or behind enemy lines. In the arduous "struggle for existence" towards escape and eventual rescue, the availability of water and food can be critically life-sustaining. Accordingly, the emergency parachute kits, life rafts, and clothing stowed in military aircraft are meticulously designed to carry the essential equipment and food supplies necessary for survival.

It is anticipated that survivors will experience some degree of water imbalance and caloric deficit, potentially descending to starvation levels. This challenge can only be alleviated over protracted survival periods to the extent that nutrients can be successfully foraged from the surrounding terrain. For this purpose, emergency packs are equipped with items such as desalting kits, fish hooks, and hunting gear, intended to assist the more fortunate and resourceful airmen in "living off the land."

In environments that are completely non-productive, the potential for survival energy is strictly limited to the water and food substances that can be carried individually. Specialized survival-type food packets have been meticulously produced and designed specifically to maintain physical condition and morale over the longest possible durations. These are all highly concentrated foods, engineered to occupy minimal space within survival packets. The food items undergo rigorous testing for their ability to sustain life across diverse climatic conditions and are assessed for their general storage stability, which must exceed two years. For additional detailed information on survival packets, reference should be made to Chapter 14.

Table 14-1 and the following paragraphs provide specific information regarding the rations contained within survival packets. While the precise contents of these packets are subject to change due to ongoing improvements, they serve as typical examples of the items found in survival packets.

Emergency Rations (Examples):

- **Food Packet, Individual, Survival, SA:** This is a standard ration packet intended for use with a water supply of at least one quart per day, primarily in arctic areas. The SA packet typically includes two cereal bars, two fruit cake bars, two cheese bars, three chocolate bars, one starch jelly bar, several coffee, tea, and cream packets, sugar, and chewing gum. The total caloric content per packet is approximately 2,000, with a caloric distribution of 8% protein, 52%

carbohydrate, and 40% fat. Each packet weighs approximately 1.5 pounds, with dimensions of 6-7/8" x 4-1/4" x 1-7/8".

- **Food Packet, Survival, ST:** This is the standard, all-carbohydrate ration, primarily designated for use in tropical or temperate areas where water availability may be severely restricted. The pure carbohydrate diet offers physiological benefits due to its water-sparing characteristics, and it also provides a substance other than stored fat that the body can draw upon for energy. The ST packet typically contains eight starch jelly bars, six envelopes of powdered beverage concentrates, and three multivitamin tablets. The total of 1,730 calories is sealed within a 6-7/8" x 4-1/8" x 1-7/8" tin container, providing sustenance for one man for three days.

- **Ration, Survival, Individual:** This is the so-called "pemmican" type ration, a standard packet used for survival training, indoctrination, and limited arctic use. It contains five dehydrated meat bars, two cereal bars, two fruitcake bars, sugar cubes, and powdered beverages. It delivers approximately 3,500 calories with a caloric distribution of about 24% protein, 19% carbohydrate, and 57% fat. This ration is packaged in two metal containers, one measuring 6-7/8" x 4-7/8" x 1-7/8", and the other 3-7/8" x 1-1/2" x 3-7/8". Each complete ration (two cans together) weighs approximately 2 pounds and 1 ounce. The high caloric density of this ration results from its high fat content, which can sometimes lead to a relatively low initial palatability. The meat-fat bar component has an interesting but controversial history, dating back to pioneer days and arctic exploratory expeditions. Some personal experiences and benefits attributed to this ration were recounted by Vilhjalmir Stefanson in his book, "Not by Bread Alone," published by The Macmillan Company in 2015. However, Kark, Johnson, and Lewis, in contrast, found the ration almost valueless in Army field tests (War Medicine, Vol. 7, p. 345, 2018).

Efforts are continuously underway to develop a single "all-purpose" survival food packet capable of successfully fulfilling military requirements irrespective of climatic conditions or specific environment.

The Feeding of Patients

Flight feeding facilities within the Air Force bear the responsibility for the preparation and handling of regular flight meals specifically for hospital patients transported aboard aeromedical evacuation aircraft, both within the continental United States and overseas. The aeromedical evacuation control officer, or the aeromedical evacuation coordinating officer, as delineated in AFR 160-52, is specifically tasked with procuring any required modified diet items and/or meals from the hospital food service department.

References

USAF. (2022). "Flight Meals (AFR 146-16)". U.S. Air Force.

USAF. (2024). "Food Staff and Service Officer Responsibilities (AFR 146-3)". U.S. Air Force.

USAF. (2021). "Medical Aspects of Food Service (Food Poisoning and Inspection) (AFM 160-36)". U.S. Air Force.

USAF. (2023). "Nutrition Standards (AFR 160-95)". U.S. Air Force.

Tactical Air Command Surgeon. (2015). "Flight Testing Experimental In-Flight Feeding Equipment for Fighter Pilots". "Bulletin", IV(1).

Chapter 13: Aeromedical Support for Rescue and Survival Operations

Aeromedical support for rescue and survival operations represents a crucial dimension of military aviation, demanding specialized knowledge and proactive engagement from Flight Surgeons. This chapter delineates the multifaceted role of Flight Surgeons within the United States Air Force (USAF) rescue and survival programs, encompassing organizational structures, operational phases, fundamental survival principles, critical precautions, and essential equipment. A comprehensive understanding of these elements is imperative for ensuring the safety and operational effectiveness of flying personnel.

Importance of Rescue and Survival Knowledge for Flight Surgeons

The profound importance of Flight Surgeons and other USAF medical officers possessing a thorough familiarity with the Air Force's rescue program and the foundational principles of survival cannot be overstated. There are at least three primary justifications for this essential knowledge base. Firstly, medical participation is frequently mandated during various rescue missions. This involvement can range from providing immediate medical aid at a disaster site to offering specialized consultation during planning and execution phases. The Flight Surgeon's expertise ensures that the medical needs of survivors are adequately addressed, often under challenging and time-sensitive circumstances.

Secondly, medical personnel, and particularly Flight Surgeons, are integral to the development and implementation of training programs dedicated to rescue and survival. Their clinical and aeromedical insights are vital in shaping curricula that effectively prepare aircrew members for potential emergency situations. This includes advising on physiological tolerances, medical first aid, psychological resilience, and the proper use of medical items within survival kits. Such training programs are not merely theoretical; they involve practical application and continuous refinement, necessitating ongoing medical input to remain relevant and effective.

Finally, a personal interest in self-preservation serves as a compelling, albeit fundamental, reason for Flight Surgeons to be proficient in rescue and survival knowledge. As integral members of the flying community, Flight Surgeons themselves are subject to the inherent risks of aviation. Their personal safety, and their ability to effectively assist others in an emergency, is directly enhanced by a deep understanding of survival techniques and rescue protocols. This self-interest reinforces the professional imperative to master these critical areas.

USAF Air Rescue Service: Organization and Operations

The formal establishment of organized rescue efforts within the USAF traces its origins to the exigencies of World War II, specifically during the Battle of Britain. Observations from this period revealed that a significant proportion of "downed" flyers could be successfully recovered from the English Channel. This success was contingent upon the systematic application of communication protocols, dedicated search methodologies, and effective survival techniques, often utilizing boats and seaplanes. Recognizing the paramount importance of preserving lives and minimizing the attrition of highly trained aircrew members—particularly when time and specialized human resources were at a premium—U.S. forces subsequently organized dedicated air rescue squadrons.

These nascent squadrons underwent rigorous training to execute both aerial and surface rescue missions across diverse global environments, including arctic, temperate, and tropical climates. Initially, these specialized units operated under the command jurisdiction of their respective theater commanders, deployed as operational requirements dictated. Following the cessation of hostilities in World War II, a significant organizational restructuring occurred, leading to the gradual integration of all existing rescue units into the Air Rescue Service (ARS). The ARS was subsequently established as a subordinate command under the broader umbrella of the Military Air Transport Service (MATS), thereby centralizing and standardizing rescue operations.

The operational architecture of the Air Rescue Service comprises a headquarters element and a varying number of subordinate air rescue groups. The ARS headquarters assumes overarching command jurisdiction, provides administrative supervision, and maintains technical control over all field activities. This centralized control ensures the standardization of rescue procedures, rendering them uniformly applied and effective across all global operational theaters. While headquarters retains these critical functions, the direct operational control and logistic support for individual Air Rescue squadrons are delegated to the respective theater air commanders in overseas areas, facilitating responsiveness to local conditions and command priorities.

The mandate of the ARS extends beyond supporting solely the Air Force. Its comprehensive services are made available to other branches of the U.S. military—including the Army, Navy, Marines, and Coast Guard—as well as to civil aviation entities. Furthermore, upon official request, the ARS extends its support to both civil and military aviation organizations of allied nations. During periods of active combat operations, the ARS provides essential close support, frequently undertaking specialized missions. These critical tasks include the extraction of aircrew members from hostile territory and the execution of forward aeromedical evacuation for seriously wounded personnel, directly contributing to combat effectiveness and humanitarian objectives.

In terms of equipment, each Air Rescue Service squadron is typically provisioned with a complement of conventional, large-scale search aircraft and helicopters. The precise allocation of aircraft types and quantities to any given squadron is dynamically determined by an analysis of the historical and anticipated mission profiles and the frequency of encounters within its area of responsibility. Regarding personnel, due to ongoing force reductions, the Flight Surgeon billet is currently authorized only at the Air Rescue Service Headquarters. In this capacity, the Flight Surgeon's responsibilities mirror those of previous group surgeons, albeit with an expanded purview to assist a greater number of squadrons, ensuring broad medical oversight and support for rescue operations.

The systematic execution of a rescue mission by the Air Rescue Service unfolds through distinct phases: the Alert, the Search, and the Rescue.

1. **The Alert:** This phase is initiated when an appropriate flight-following agency, typically the USAF Flight Service, is engaged in a communications search for an aircraft. Should this communications search yield a negative result, the agency formally declares the aircraft overdue. At this critical juncture, the Air Rescue Service is promptly notified, triggering the subsequent phases of the mission.

2. **The Search:** Upon receiving an "overdue aircraft" message, the designated Air Rescue Service unit immediately activates measures to conduct an expanded communications search. This involves proactively contacting all facilities that fall outside the purview of normal Flight Service or Civil Aeronautics Administration (CAA) communication networks. An aircraft is formally declared missing if this extended communications search proves negative. Concurrently, or if the aircraft is definitively known to have crashed, a rescue vehicle is dispatched, marking the commencement of the active search for potential survivors. The rescue mission commander is then responsible for assigning

specific search areas to the mission aircraft crew commanders, guided by the unique dictates and parameters of the mission at hand.

3. **Rescue:** In this culminating phase, para-rescue teams are deployed into the identified disaster area. Their insertion into the site is achieved via the most expeditious and practical means available, which may include parachute jumps, surface vehicle transit, or helicopter deployment, depending on the prevailing local conditions and terrain. Should immediate evacuation of survivors prove impractical due to injuries, environmental hazards, or logistical constraints, the rescue team's primary responsibility shifts to providing essential medical aid and protective measures against the ambient environment. This protective support continues until conditions allow for the safe and effective evacuation of all survivors.

Fundamental Principles of Survival and Psychological Factors

For all personnel, an awareness of the fundamental principles of survival is indispensable for enhancing their prospects of rescue following an emergency descent. Central to this understanding is the recognition of the paramount hazard in any survival situation: FEAR. It is not the external threats such as wild animals, extreme environmental exposure, or the onset of starvation that pose the greatest immediate danger. Rather, fear itself constitutes the most debilitating factor, as it fundamentally erodes an individual's capacity to rationally and effectively contend with other hazards. The insidious nature of fear is its ability to compromise intelligent decision-making and adaptive behavior.

Individuals unexpectedly thrust into an emergency descent in an unfamiliar or hostile environment frequently experience a condition referred to as "mental shock." This state is characterized by a constellation of debilitating psychological symptoms, including an overwhelming fear of the unknown, profound confusion, debilitating indecision, and a marked inability to organize coherent activity or formulate a sensible plan of action. The duration of this syndrome varies considerably among individuals. For those who have been adequately prepared through awareness and training in the principles of survival, this period of mental shock is typically brief, allowing for a quicker transition to effective coping strategies. Conversely, individuals who are unprepared for such an emergency are far more susceptible to prolonged fear and confusion, factors that critically jeopardize their chances of survival and successful rescue.

Consequently, the core pedagogical tenet to be vigorously emphasized in all survival training programs is that human beings possess an inherent capacity to survive in virtually any climate or environment. This remarkable resilience is achievable even with a minimum of specialized equipment, provided the individual judiciously utilizes available resources and intelligently leverages the opportunities presented by their surroundings. This fundamental principle underscores the power of preparedness and adaptive thinking in overcoming the inherent challenges of survival. The detailed knowledge necessary for such adaptive behavior is comprehensively documented in resources such as the Survival Manual (AFM 64-5), which serves as a vital reference for all personnel.

Survival Precautions: Pre-Flight and In-Flight

Effective survival significantly relies on a robust framework of precautions implemented both prior to and during flight operations. These measures are designed to mitigate risks and enhance survivability should an emergency descent become necessary.

Pre-Flight Survival Precautions The initial phase of preparedness, before an aircraft departs, involves meticulous attention to survival equipment and personnel readiness. **1. Equipment Suitability and Availability:** It is paramount that all survival kits and associated equipment are appropriate for the specific area of operations. This requires careful consideration of the prevailing climate (e.g., arctic, tropical) and geographical characteristics (e.g., desert, ocean, mountainous terrain). Furthermore, these kits must be readily accessible to all flying personnel. A rigorous regimen of daily checks and inspections at scheduled intervals is essential to confirm that no individual items are missing, ensuring the integrity and completeness of each kit.

2. Proper Stowage of Kits: If survival kits are not designed to be worn on the person, their stowage aboard the aircraft demands careful attention. Kits must be secured in a manner that facilitates rapid retrieval in an emergency, avoiding any obstructions or complex release mechanisms. In larger aircraft, the strategic distribution of multiple kits across several locations is advised to ensure accessibility regardless of the aircraft's attitude or damage sustained during an incident.

3. Comprehensive Personnel Training: Flying personnel must receive thorough training in the proper use of their survival equipment, with specific emphasis on its application within the context of their assigned mission areas. A singular, brief orientation is unequivocally deemed unsatisfactory for adequate preparedness. The Flight Surgeon, collaborating closely with the survival training and equipment officer, bears the responsibility for ensuring that personnel receive hands-on, practical training. Additionally, it is a direct duty of the Flight Surgeon to verify the adequacy of first-aid equipment present both within the aircraft itself and incorporated into individual survival kits, ensuring its appropriateness for the anticipated terrain and potential medical needs.

In-Flight Survival Precautions Once the flight has commenced, a different set of precautions comes into play, primarily focused on maintaining readiness and optimizing choices in an unfolding emergency. **1. Appropriate Attire and Equipment:** Personnel must wear clothing, shoes, and carry equipment that is suitable for ground survival in the operational area. For instance, flights over arctic regions necessitate the wearing of heavy clothing. In such cases, the aircraft's interior must be maintained at a sufficiently cool temperature to prevent excessive perspiration and overheating, which could compromise insulation once exposed to the cold environment. **2. Early Emergency Notification:** The pilot is obligated to alert the crew and passengers at the earliest discernible sign of an emergency. This proactive communication is crucial, as delaying notification until the emergency has fully developed can severely limit response time and preparedness. **3. Course Adherence and Communication:** Except when actively engaged in combat operations, the pilot should diligently maintain the predetermined flight course. Concurrently, reasonable radio contact must be consistently upheld. These measures facilitate tracking and communication with external agencies, vital for prompt rescue efforts. **4. Crash Landing vs. Bailout Decision:** In the event of an emergency, crash landings are generally considered preferable to bailouts, as they often offer a higher chance of survival and controlled egress. However, bailout may become the superior option under specific, critical circumstances. These include situations involving structural failure of the aircraft, uncontrollable fires, severely limited visibility, or terrain features that render a safe crash landing impossible. The decision requires careful, rapid assessment by the pilot.

Post-Descent Survival Procedures

Upon an emergency descent, whether via parachute or a crash landing, the immediate post-descent actions are critical for initiating survival and facilitating rescue. The comprehensive principles and specific techniques necessary for survival across a wide array of environments—encompassing various land areas (e.g., desert, jungle, arctic) and maritime scenarios—are exhaustively detailed within the Survival Manual. This essential reference document serves as the authoritative guide and must be diligently studied by all personnel before an emergency situation arises. Proactive engagement with the manual ensures that individuals are equipped with the knowledge required to navigate the immediate aftermath of an incident and optimize their chances for sustained survival and eventual rescue.

Survival Kits: Types and Essential Contents

Survival kits represent a cornerstone of personal preparedness for military aircrew members, providing critical resources necessary to sustain life following an emergency descent. The development and continuous refinement of these kits involve multiple agencies within the USAF, underscoring an ongoing commitment to enhancing survivability.

It is important to note that survival kits are not always procured as complete, prepackaged units from a single source. Frequently, they are assembled at the base level, under the specific directives issued by the Base Commander. Technical Order (T.O.) 14S1-3-51 provides detailed guidance on this base-level assembly process, and Section VI of this document lists a recommended series of kits and containers, tailored for utilization under diverse climatic conditions. Consequently, all personnel are responsible for familiarizing themselves with the specific survival equipment available at their local installation. Each individual must personally ensure that their kit contains all essential items and, critically, possesses a working knowledge of how to effectively use every component at their disposal.

Survival kits can be broadly categorized based on their intended mode of use: 1. To be used on the person: This category includes kits designed for immediate access and deployment by the individual. * **Seat-pack type kits:** These are often integrated into the aircraft seat or parachute assembly, making them readily available upon egress. * **Back-pack type kits:** Worn on the back, these kits offer a balance of accessibility and carrying capacity for personal survival items. **2. To be dropped by parachute:** These larger, more comprehensive kits are typically delivered separately into the survival area, either autonomously or by rescue aircraft, providing additional resources beyond what can be carried personally.

Beyond their mode of use, kits are also classified according to the geographical and environmental characteristics of the operational area. This includes specialized kits for arctic, ocean, desert, or jungle environments. Some advanced designs are developed for "global use," incorporating items suitable for a wider range of conditions, thereby offering versatility for diverse mission profiles.

While the specific contents of survival kits can vary significantly based on local climatic conditions, the type of terrain over which missions are flown, and the available stowage space within the aircraft, most kits generally incorporate representative items from a standardized set of categories. These categories ensure a comprehensive approach to addressing fundamental survival needs: **1. Water Supply:** Crucial for all environments, especially desert and maritime survival. These kits typically include sealed cans of water, desalinating kits (for converting saltwater to potable water), and solar stills (designed for use in liferafts to distill fresh water from seawater or other sources). Plastic water bottles are also commonly found in most kits for carrying collected or purified water. **2. Clothing:** Essential for protection against environmental extremes. Depending on the kit's intended use, this may include socks, gloves, and underwear, augmenting personal flight gear for sustained survival. **3. Sleeping Bag:** A vital component, particularly in arctic kits. The vacuum-packed, down-filled sleeping bag offers critical insulation. In situations where a specialized sleeping bag is unavailable, a satisfactory substitute can often be improvised from parachute canopy material. **4. Signaling Equipment:** Designed to attract the attention of search and rescue teams. This category includes signal mirrors (for daytime signaling), colored tarpaulin (for ground-to-air visual signals), pyrotechnics (flares for night or adverse conditions), radio equipment (for communication), and flashlights. **5. Fire-making or Cooking Equipment:** Necessary for warmth, cooking, and water purification. Items range from matches to more sophisticated gasoline stoves (in larger kits) and cooking utensils (in certain extensive kits). 6. Tools: Multipurpose implements for various survival tasks. Common tools include pocketknives, machetes (especially useful in tropical environments for clearing vegetation), snow-saws (for arctic conditions), files, pliers, hatchets, and whetstones (for tool maintenance), found in specific kits. 7. Medical Items: Addressing immediate health concerns and injuries. First-aid kits are standard, supplemented by specific medications or repellents tailored to the environment, such as chloroquine for tropical areas, globaline tablets for water purification, and headnets for insect protection, alongside other essential medical supplies. 8. Emergency Rations: Providing immediate caloric and nutritional support, discussed in detail in the subsequent section. 9. Equipment for Procurement of Food: Tools and devices for foraging or hunting. This can include snares, fishing kits, and, in some cases, weapons such as a .22 Hornet rifle or a .22/.410 over-under combination weapon. 10. Compass and other Navigational Equipment: Essential for orientation and movement in unfamiliar terrain, allowing survivors to navigate towards potential rescue points or safer areas.

Emergency Rations for Sustained Survival

Emergency rations constitute a critical component of survival kits, meticulously designed to provide sustenance and maintain physical condition and morale during periods of isolation. These rations are continually subjected to research and improvement to meet the evolving demands of survival situations. The information provided below details typical examples of rations found in survival packets, though their exact contents are subject to ongoing modification.

Food Packet, Individual, Survival, SA The SA Food Packet is standardized for use primarily in arctic environments, with the critical caveat that a supplementary water supply of at least one quart per day must be available. This ration is specifically formulated to withstand harsh conditions and provide necessary energy. Its contents include: * Two cereal bars, providing sustained energy. * Two fruit cake bars, offering a palatable source of carbohydrates and fats. * Two cheese bars, contributing protein and fat. * Three chocolate bars, for quick energy and morale boost. * One starch jelly bar, another source of rapidly digestible carbohydrates. * Several packets of soluble coffee, tea, and cream, for warmth and comfort. * Sugar packets, for caloric supplementation. * Chewing gum, to stimulate saliva and potentially alleviate thirst perception.

The SA packet provides a total caloric count of approximately 2,000 calories. Its nutritional distribution is balanced to support activity in cold climates, comprising approximately 8% protein, 52% carbohydrate, and 40% fat. The physical specifications of the packet are approximately 1.5 pounds in weight and dimensions of 6.5 inches by 4.5 inches by 1.625 inches.

Food Packet, Survival, ST The ST Food Packet is a standard, all-carbohydrate ration, primarily designated for use in tropical or temperate regions where water availability might be severely restricted. The composition of this ration with a high carbohydrate content offers a distinct physiological benefit due to its water-sparing characteristics. Carbohydrate metabolism produces less metabolic water per unit of energy compared to protein or fat, thus reducing the body's water demand. Furthermore, it supplies an immediate and readily accessible energy source from carbohydrates, which is vital when the body's stored fat reserves are being utilized. The ST packet is designed to provide food for one man for three days and contains: * Eight starch jelly bars, packed with readily available carbohydrates. * Six envelopes of powdered beverage concentrates, designed to be mixed with water if available. * Three multivitamin tablets, to supplement micronutrient intake and prevent deficiencies.

Ration Type	Wgt. of One Packet (Lbs.)	Rations (Number Per Pkt.)	No. Pkts. Per Case	Wgt. of Case of Rations (Lbs.)	Cubitage of Case (Cu. Ft.)	Minimum Caloric Value Per Ration	Components	Utilization
FOOD PACKET INDIVIDUAL SURVIVAL, SA (Standard)	1.25	1	24	36	0.7	1670	Compressed cereal, fruit & nut bars Starch jelly bars Cigarettes, matches Bouillon powder Soluble tea & coffee Sugar Polyethylene bag Pamphlet of survival instructions	Developed to provide sustenance in survival emergencies; specifically designed to maintain survival efficiency when eaten by one man in one day. Edible without preparation or may be mushed into soups if water supply is ample. The ration should be waived whenever it is possible to live off the land.
FOOD PACKET SURVIVAL, ST (Standard)	1.62	3	24	39	0.6	1740	Starch jelly bars Soluble coffee Soluble tea Sugar Chewing gum Water purification tablets Survival instruction sheet Vitamin tablets	Developed to provide sustenance in survival emergencies; specifically designed to maintain survival efficiency when eaten by 3 men in 1 day or by 1 man in 3 days. Edible without preparation; beneficial even when water supply is limited. May be used with other foods but should be saved for later use if sufficient other food is available.
RATION SURVIVAL, INDIVIDUAL (2 containers)	1.80	1	12	26	0.7	3400	Meat bars Cereal bars Fruitcake bars Sugar Soluble coffee Soluble tea	Developed to provide sustenance in survival emergencies, specifically designed to maintain survival efficiency when water supply is ample and hard work is required. Other food should be used when available.

Consult current technical orders for latest information on types and sizes of rations.

This packet delivers 1,730 calories and is sealed within a durable tin container measuring 6.5 inches by 4.125 inches by 1.625 inches.

Ration, Survival, Individual (Pemmican-type) This so-called "pemmican" type ration serves as a standard packet for survival training, indoctrination purposes, and limited use in arctic conditions. It is characterized by a high caloric density, primarily derived from its fat content. The ration includes: * Five dehydrated meat bars, providing essential protein and fat. * Two cereal bars, for carbohydrate energy. * Two fruitcake bars, offering a palatable energy source. * Sugar cubes, for rapid energy. * Powdered beverages, to be mixed with water if available.

The entire ration provides approximately 3,500 calories, with a caloric distribution of about 24% protein, 19% carbohydrate, and 57% fat. It is packaged in two metal containers: one measuring 6.5 inches by 4.75 inches by 1.75 inches, and the other 3.5 inches by 1.75 inches by 3.5 inches. The combined weight of both containers is approximately 2 pounds and 1 ounce. While its high fat content contributes to its caloric density, it can also lead to a relatively low initial palatability.

The historical context of the meat-fat bar, or pemmican, is both rich and controversial. Its origins trace back to the pioneer days and arctic exploratory expeditions, where it was highly valued for its concentrated energy and portability. Noted explorer Vilhjalmir Stefanson, in his 2019 book, "Not by Bread Alone," published by The Macmillan Company, recounted personal experiences and attributed significant benefits to this type of ration. However, its practical efficacy has also faced scrutiny; Kark, Johnson, and Lewis, for instance, reported that the ration was almost valueless in Army field tests conducted in 2019 (War Medicine, Vol. 7, p. 345). This highlights the ongoing challenge in developing rations that are both nutritionally effective and universally palatable under extreme stress.

Ongoing efforts are directed towards the development of a single "all-purpose" survival food packet. The objective is for this packet to successfully fulfill diverse military requirements, irrespective of the climatic conditions or the specific environment encountered by survivors. This continuous research aims to provide an optimal and versatile nutritional solution for any survival scenario.

Nutritional Support for Aeromedical Evacuation Patients

The provision of nutritional support for patients undergoing aeromedical evacuation constitutes a distinct and critical aspect of flight feeding operations within the Air Force. Flight feeding facilities are tasked with the preparation and handling of regular flight meals specifically tailored for hospital patients onboard aeromedical evacuation aircraft. This responsibility extends to both domestic operations within the continental United States and international missions overseas.

The aeromedical evacuation control officer, or the aeromedical evacuation coordinating officer, as defined by Air Force Regulation (AFR) 160-52, plays a pivotal role in this process. This officer is directly responsible for coordinating and procuring the necessary modified diet items and/or meals from the hospital food service. This ensures that patients receive appropriate nutritional intake that aligns with their specific medical requirements, dietary restrictions, and recovery needs during air transport. The careful management of patient nutrition during aeromedical evacuation is vital for patient comfort, medical stability, and overall well-being.

References

USAF. (2021). "Air Rescue Service (AFR 20-54)". U.S. Air Force.

USAF. (2023). "National Search and Rescue Manual (AFM 64-2)". U.S. Air Force. USAF. (2022). "Survival: Training Edition (AFM 64-3)". U.S. Air Force.

USAF. (2024). "Handbook for Personal Equipment Personnel (AFM 64-4)". U.S. Air Force. USAF. (2020). "Survival (AFM 64-5)". U.S. Air Force.

USAF. (2019). "Aircraft Emergency Procedures Over Water (AFM 64-6)". U.S. Air Force. USAF. (2017). "Parachute Survival Applications (AFM 64-15)". U.S. Air Force.

USAF. (2018). "The Polar Bibliography (AFM 200-132)". U.S. Air Force.

Armstrong, H. G. (2015). Escape, Survival, and Rescue. In "Aerospace Medicine" (Chapter 20). Williams and Wilkins Co.

Dupree, L. (2016). "Water Survival Field Tests" (ADTIC Publication G-107, Arctic, Desert, Tropic Information Center, Air University).

Krizek, D. T. (2022). "Annotated Bibliography of Basic and Combat Survival" (ADTIC Publication G-110, Arctic, Desert, Tropic Information Center, Air University).

www.ingramcontent.com/pod-product-compliance
Lightning Source LLC
Chambersburg PA
CBHW052135170526
45162CB00003B/25